Fundamental and Applied Nano-Electromagnetics II

NATO Science for Peace and Security Series

This Series presents the results of scientific meetings supported under the NATO Programme: Science for Peace and Security (SPS).

The NATO SPS Programme supports meetings in the following Key Priority areas: (1) Defence Against Terrorism; (2) Countering other Threats to Security and (3) NATO, Partner and Mediterranean Dialogue Country Priorities. The types of meetings supported are generally "Advanced Study Institutes" and "Advanced Research Workshops". The NATO SPS Series collects together the results of these meetings. The meetings are co-organized by scientists from NATO countries and scientists from NATO's "Partner" or "Mediterranean Dialogue" countries. The observations and recommendations made at the meetings, as well as the contents of the volumes in the Series, reflect those of participants and contributors only; they should not necessarily be regarded as reflecting NATO views or policy.

Advanced Study Institutes (ASI) are high-level tutorial courses to convey the latest developments in a subject to an advanced-level audience.

Advanced Research Workshops (ARW) are expert meetings where an intense but informal exchange of views at the frontiers of a subject aims at identifying directions for future action.

Following a transformation of the programme in 2006, the Series has been re-named and re-organised. Recent volumes on topics not related to security, which result from meetings supported under the programme earlier, may be found in the NATO Science Series.

The Series is published by IOS Press, Amsterdam, and Springer, Dordrecht, in conjunction with the NATO Emerging Security Challenges Division.

Sub-Series

A.	Chemistry and Biology	Springer
B.	Physics and Biophysics	Springer
C.	Environmental Security	Springer
D.	Information and Communication Security	IOS Press
E.	Human and Societal Dynamics	IOS Press

http://www.nato.int/science
http://www.springer.com
http://www.iospress.nl

Series B: Physics and Biophysics

Fundamental and Applied Nano-Electromagnetics II

THz Circuits, Materials, Devices

edited by

Antonio Maffucci

Department of Electrical and Information Engineering
University of Cassino and Southern Lazio, Cassino, Italy

and

Sergey A. Maksimenko

Institute for Nuclear Problems
Belarusian State University, Minsk, Belarus

Springer

Published in Cooperation with NATO Emerging Security Challenges Division

Proceedings of the NATO Advanced Research Workshop on Fundamental and Applied NanoElectroMagnetics II: THz Circuits, Materials, Devices Minsk, Belarus
5–7 June 2018

ISBN 978-94-024-1689-3 (PB)
ISBN 978-94-024-1689-3 (HB)
ISBN 978-94-024-1687-9 (e-book)
https://doi.org/10.1007/978-94-024-1687-9

Published by Springer,
P.O. Box 17, 3300 AA Dordrecht, The Netherlands.

www.springer.com

Printed on acid-free paper

Preface

The increasingly intensive penetration of nanotechnologies has led to the birth of the "nanoelectromagnetics," a novel applied science that studies the interaction of electromagnetic radiation with quantum mechanical low-dimensional systems. The growing interest to nanoelectromagnetics is due to the impressive progress of nanotechnology in the last decades, which nowadays allows us to employ in electronic and photonic devices' novel nanoscale materials such as nanostructured carbon, noble metal nanowires, semiconductor heterostructures, quantum dots, and organic macromolecules.

The route from theoretical principles to technology applications requires interdisciplinary know-how and expertise in the fields of fundamental and applied electromagnetics, chemistry and technology of nanostructures and nanocomposites, and physics of nanostructured systems. Indeed, the workshop on "Fundamental and Applied Nanoelectromagnetics" held in Minsk, Belarus, from 5 to 7 June 2018, had the main scope to offer a meeting opportunity to the interdisciplinary community of researches that is nowadays involved in nanoelectromagnetics.

The workshop was jointly organized by the Department of Electrical and Information Engineering of the University of Cassino and Southern Lazio, Cassino, Italy, and by the Institute for Nuclear Problems of Belarusian State University, Minsk, Belarus, thanks to the fundamental support from NATO Science Committee, which awarded the workshop with a grant, under the NATO Science for Peace and Security Program.

This book is the result of the above workshop that specifically addressed topics related to the Terahertz technology. The reader will find here a comprehensive overview on the synthesis of different nanostructured materials, study of their electrical and optical properties, use of nano-sized elements and nanostructures as building blocks of devices, and design and fabrication of nanotechnology devices operating in the THz, IR, and optical range.

The book introduces the reader to materials like nanocomposites, graphene nanoplatelets, carbon nanotubes, metal nanotubes, and silicon nanostructures; to devices like photonic crystals, microcavities, antennas, and interconnects; and to applications like sensing and imaging, with special emphasis on the THz frequency range.

We are glad to acknowledge all the authors for their high-level contributions.

Cassino, Italy — Antonio Maffucci
Minsk, Belarus — Sergey A. Maksimenko

Contents

Chapter 1
Electromagnetic Response of Carbon Nanotube-Based Composites

Mikhail V. Shuba

Abstract We analyze electromagnetic parameters of carbon nanotube (CNT) based composite. The simple theory for low-content composites comprising non-interacting CNTs can provide the main physical mechanisms determining their electromagnetic response. Frequency and concentration dependencies of the effective conductivity of the composites in the microwave and terahertz ranges are discussed.

Keywords Carbon nanotubes · Composite · Microwave · Terahertz · Effective conductivity

1.1 Introduction

Single-walled carbon nanotube (SWCNT) can be considered as graphene sheet rolled up to form a seamless cylinder [1, 2]. Depending on the direction of rolling, it can be metallic or semiconducting. SWCNT is characterized by chiral indexes (n, m) and it has a typical radius of 0.8–2 nm and length 10^2–10^6 nm. Multiwalled carbon nanotube (MWCNT) comprises concentric shells ($\lesssim$or tubes). Due to the periodical boundary condition for electron wave function in the circumference direction, electrons move along the tube axis with permanent angular momentum demonstrating 1D electron transport [3–5]. The variation of the electron angular momentum is forbidden by the selection rules resulting in the excitation of only axially-symmetrical current density in carbon nanotube (CNT).

In this report we shall consider a frequency range below interband electron transitions, i.e. below 10 THz. In this range, as shown from quantum-mechanical consideration [3, 4], the axial surface electrical conductivity σ_{cn} of the metallic SWCNT of small diameter (<2 nm) follows Drude law

M. V. Shuba (✉)
Institute for Nuclear Problem, Belarusian State University, Minsk, Belarus
e-mail: mikhail.shuba@gmail.com

A. Maffucci, S. A. Maksimenko (eds.), *Fundamental and Applied Nano-Electromagnetics II*, NATO Science for Peace and Security Series B: Physics and Biophysics, https://doi.org/10.1007/978-94-024-1687-9_1

$$\sigma_{cn}(\omega) = \frac{2ie^2 \upsilon_F}{\pi \hbar R_{cn}(\omega + i\nu)} \tag{1.1}$$

where υ_F is the velocity of electron on the Fermi level, $\upsilon_F \approx 10^6$ m/c, $\nu = 1/\tau$ is the relaxation frequency, τ is the relaxation time of electrons, e is the electron's charge, R_{cn} is a radius of nanotube, $\hbar$ is the reduced Planck constant, $\omega = 2\pi f$ is an angular frequency, and f is a frequency.

Semiconducting CNTs have an electronic band gap 0.5–1 eV [6]. The theory predicts a very small low frequency conductivity of the semiconducting CNTs in comparison with the conductivity of metallic tubes [3]. However, as found in experiment [7, 8], the dc and terahertz conductivity of a film comprising mostly semiconducting tubes is only several times less than the conductivity of the film containing mostly metallic tubes. The explanation could be that semiconducting tubes are unintentionally doped. Doping leads to a shift of the Fermi level resulting in partial or complete "metallization" of semiconducting tubes [9].

The theory of electromagnetic (EM) wave scattering has been developed for individual SWCNT [10–12], multiwalled CNT [13], SWCNT bundle [14], SWCNT with dielectric coating [15], curved SWCNT [16] and SWCNT with mesoscopic insertion [17]. Moreover, the EM wave interaction with 2D array of SWCNTs [18–20] was also presented. The modeling of dc-conductivity of CNT network can be found in [21, 22]. However, to the best of our knowledge, there is no theory describing frequency dependence of the complex effective permittivity of the CNT-based composite in the range from 0 Hz up to 1 THz. This theory, first of all, should take into account the intertube electron tunneling. At present, the simple Waterman-Truell model [23] of non-interacting nanoparticles has been applied to describe qualitatively the behavior of the effective conductivity in the microwave and terahertz ranges [24, 25].

Collective electron excitations coupled with EM field are called one-dimensional plasmon [26] or surface wave [4]. Hereinafter we will call them as surface wave. In SWCNTs, it has very short wavelength λ_p due to very low phase velocity $\upsilon_p \approx 0.01c - 0.05c$, where c is speed of light in vacuum. The formation of the standing surface wave in finite-length CNT at $L \approx n\lambda_p/2$, $n = 1, 2\ldots$ leads to resonances in the spectrum of CNT polarisability [11]. They are called geometrical resonances or antenna resonances [11]. The first antenna resonance ($n = 1$) is called as localized plasmon resonance (LPR) [27]. Odd antenna resonances can be excited by the plane wave [11], whereas all the resonances occur in the spectra of the thermal radiation of CNTs [28]. The frequency of the LPR f_p increases from 3 THz up to 30 THz when the tube length decreases from 1000 to 100 nm.

Frequency f_p divides the spectral range into three regimes of CNT interaction with the EM field [25]: quasi-static ($f << f_p$), resonant ($f \approx f_p$), and dynamic ($f > f_p$) regimes. In order to describe these regimes in details, let us consider a CNT being oriented along z-axis and exposed to incident alternative field $E = E^0 e^{i\omega t}$ directed along z-axis, t is a time. The axial surface current density $j(z)$ induced on the CNT surface determines the EM response of CNT. It is calculated from Ohm's law

$$j(z) = \sigma_{cn} E_z^{tot}(z). \tag{1.2}$$

Axial component of the total field on the CNT surface $E_z^{tot}(z)$ is a sum of the incident $E_z^0(z)$ and scattered $E_z^{sc}(z)$ fields

$$E_z^{tot}(z) = E_z^0(z) + E_z^{sc}(z). \tag{1.3}$$

In the quasi-static regime, due to the polarization effect, the unlike charges concentrate on the opposite edges of the CNT. These charges create the scattering field on the surface of the CNT with axial $E_z^{sc}(z)$ and radial $E_\rho^{sc}(z)$ components: $E_z^{sc}(z) \approx -E_z^0(z)$ and $E_\rho^{sc}(z) >> E_z^0(z)$. The former relation leads to the screening effect in the CNT [15, 25] providing the inequality $E_z^{tot}(z) << E_z^0(z)$, while the latter relation results in the near field enhancement effect, $E_\rho^{tot}(z) >> E_z^0(z)$ [15, 29, 30]. Screening effect is the main mechanism resulting in a weak interaction of EM radiation with CNTs. Screening effect is stronger for shorter CNTs. It is also stronger at lower frequencies. The influence of the screening effect on the effective permittivity of the CNT-based composite is discussed in Sect. 1.3.

In the quasi-static regime, carbon nanotube has a high value of the real part and low value of the imaginary part of the polarizability α, i.e. $\mathrm{Re}(\alpha) > > \mathrm{Im}\,(\alpha)$. This property can be utilized for the fabrication of a high dielectric constant and low loss material in the terahertz range. It has been shown in [31], that at a frequency of 0.75 THz, the films comprising short-length (100–300 nm) SWCNTs or MWCNTs have effective relative permittivity with a large real part (25–136) at relatively small dielectric loss tangent (0.35–0.6). The theoretical consideration predicts that the composite material comprising randomly oriented individual CNTs and electrically isolated from each other may have the real part of the permittivity to be as much as 300 and the dielectric loss tangent to be as low as 0.03 at 20% CNT volume fraction in the terahertz range [31]. Alignment of the tubes may increase a real part of the permittivity up to the value of 900 in the direction of the tube axis and decrease one down to unity in the perpendicular direction [32]. This will be useful for the fabrication of high dielectric constant anisotropic materials for the terahertz range.

Resonant regime of CNT interaction with the terahertz field, $f \approx f_p$, can be useful for terahertz nanoplasmonics [33] and in the terahertz metamaterial area [34]. In this regime $E_\rho^{tot}(z) >> E_z^0(z)$, so that CNT can concentrate EM energy in a small volume and significantly enhance the EM field in the near-field zone [29]. Experimental evidence of the LPR in CNTs has been given in [27, 36]. It has been shown in [27, 35–37] that resonance frequency f_p depends on carbon nanotube length.

In the *dynamic regime*, $E_z^{tot}(z) \approx E_z^0(z)$, the current density excited practically does not depend on the tube length.

By variation of the tube length, it is possible to shift the frequency f_p and, consequently, to change a regime of the CNT interaction with the EM field. In practice, two regimes (quasistatic and dynamic) can occur simultaneously in CNT-based composites. Some fraction of CNTs is dispersed in a dielectric matrix in a way that they are electrically isolated from adjacent tubes. These CNTs interact with the EM radiation in quasi-static regime and contribute mostly into the real part of the effective permittivity of the composite. Other tubes are dispersed in a way that they form a conductive network. Due to the intertube current, the charges no longer concentrate or slightly concentrate near CNT edges, so that the depolarizing field does not compensate the incident field. As a result the screening effect is small and the dynamic regime occurs for CNTs within the conductive network. Those tubes contribute mostly into the imaginary part of the effective permittivity of the composite. CNTs can be also aggregated forming dense agglomerates. The effective permittivity of the agglomerates is very high (both the real and imaginary parts are hundreds or even thousands in magnitude). The incident EM field penetrates slightly into the agglomerates due to the screening effect. Thus, the aggregation effect prevents effective interaction of CNTs with the EM field.

Since CNT length L is much less than the EM wavelength λ, individual nanotube can be considered in the EM theory as electrically small nanoparticles. It has the same scattering pattern as a dipole wire antenna [11]. The scattering cross section of the CNT is much less than its absorption cross section [11, 29]. EM interaction between non-parallel tubes has not been studded yet. However, it is expected to be small as the total field along each tubes mostly depends on the charge distribution on its surface and slightly depends on the scattering field created by adjacent tubes. The EM interaction between parallel tube is quite strong [12, 14]. As shown in [14], the bundle comprising parallel metallic tubes has a set of the axially symmetric surface waves. The phase velocity υ_F of these waves in thin budles is $\upsilon_F \approx 0.01c$, it increases with the number of conductive tubes in the bundle. The frequency of the first antenna resonance f_p for each mode increases with the number of metallic CNTs in the bundle [14]. It is impossible to shift f_p into the near-infrared range as the interband transitions in CNTs lead to the strong attenuation of the surface wave [13] preventing antenna resonances in this range.

It was also shown theoretically [13] that if all the shells are metallic, the surface wave can propagate in MWCNT. The LPR has been detected in the terahertz range (0.1–1.0 THz) for thick MWCNT with relatively perfect crystalline structure [38]. However, normally the LPR is not observed in MWCNTs owing to interband transitions in semiconducting shells. However, substitutional doping may lead to the Fermi level shift and, consequently, to "metallization" of the semiconducting shells. It has been shown [39] that the LPR phenomena can exist in boron doped MWCNTs in the middle infrared range.

Thus, such phenomena as the localized plasmon resonance, the screening and near field enhancement effects determine the EM response of CNTs within the composite materials. This Chapter is providing contributions to theoretical and experimental investigations of the frequency and concentration dependences of the effective conductivity of CNT-based composites.

1.2 Theoretical Approach

The complete theory of the interaction of the EM field with CNT network still does not exist. In order to describe the EM parameters detected for CNT-based composites in a wide range from radio-frequency up to terahertz range, for simplicity, we shall consider the composite with randomly dispersed and randomly oriented CNTs with negligible mutual interaction. Further theoretical description will be based on the Waterman-Truell formula for the effective permittivity of the composite comprising individual inclusions [24]:

$$\varepsilon_{eff}(f) = \varepsilon_h(f) + \frac{1}{3\varepsilon_0}\sum_m \int_0^{\infty} \alpha_m(f, L)\, N_m(L) dL \tag{1.4}$$

where $\varepsilon_h(f)$ is relative permittivity of the host media; $\varepsilon_0 = 8.85 \cdot 10^{-12}$ F/m; $\alpha_m(f, L)$ is axial polarisability of the inclusion of type m and length L, inclusions could be SWCNTs, MWCNTs or SWCNT bundles; the function $N_m(L)$ describes the number density of inclusions; the factor 1/3 in Eq. (1.4) is due to the random orientations of the inclusions. The axial polarizability of isolated SWCNTs can be computed as follows

$$\alpha_m(f, L) = \frac{R_m}{f E_z^{(0)}} \int_0^{L} j(z) dz, \tag{1.5}$$

where $j(z)$ is the axial surface current on the SWCNT directed along axis z and located in the interval $z \in [0, L]$, R_m is a radius of SWCNT of a type m. $j(z)$ can be calculated using the integral equation technique where SWCNT is modeled as a hollow cylinder with infinitely thin walls and the effective axial surface conductivity computed from the quantum mechanical approach [10, 11]. Integral-equation technique has been also developed for MWCNT [13], bundle of SWCNTs [14], SWCNT covered by dielectric coating [15], and for SWCNT with mesoscopic insertion [17].

The effective conductivity of the composite is as given

$$\sigma_m(f) = 2\pi f \varepsilon_0 \varepsilon_{eff}(f). \tag{1.6}$$

SWCNT can be also modeled as a prolate spheroid made of an orthorhombic material with relative permittivity in the axial direction [24].

$$\varepsilon_{cn}^{(m)} = 1 - \frac{\left(\omega_p^{(m)}\right)^2}{\omega(\omega + i\nu)}, \tag{1.7}$$

where

$$\omega_p^{(m)} = \frac{e}{R_m \pi \hbar}\sqrt{\frac{6\gamma_0 b}{\varepsilon_0}}. \tag{1.8}$$

Here $\gamma_0 = 3\,\text{eV}$ is the overlap integral; R_m is a radius of SWCNT, $b = 0.142$ nm is the interatomic distance in graphene. The relative permittivity in any direction normal to the axis equals unity. Then, the axial polarizability of SWCNT is calculated as follows

$$\alpha_m = -\frac{LR_j^2}{6}\frac{\left(\omega_p^{(m)}\right)^2}{\omega^2 - \left(\omega_p^{(m)}\right)^2 N^{(m)} + i\omega\nu}, \tag{1.9}$$

where the depolarizing factor $N^{(m)}$ is determined by geometric parameters of the SWCNT,

$$N^{(m)} = 4(R_m/L)^2\left[\ln\left(L_m/R_j\right) - 1\right]. \tag{1.10}$$

In view of (1.7), after substitution (1.9) into (1.4) we arrive at

$$\varepsilon_{eff} = \varepsilon_h\left(1 + \sum_j \frac{F^{(m)}}{3}\frac{\varepsilon_{cn}^{(m)} - \varepsilon_h}{\varepsilon_h + N^{(m)}\left(\varepsilon_{cn}^{(m)} - \varepsilon_h\right)}\right), \tag{1.11}$$

where $F^{(j)}$ is the volume fraction occupied by the CNTs (conceived as ellipsoids of volume $2\pi R_j^2 L_j/3$).

As follows from (1.9) and (1.10), the axial polarizability of the metallic tubes depends slightly on the nanotube radius [24]. The polarizability of individual metallic SWCNT or bundle comprising tenth of tubes in the quasi-static regime are close to the polarizability of infinitely conductive nanorode of the same size [30]. This means that the EM response in the far zone is practically the same both for individual SWCNT and bundle comprising tenths of the tubes. However, number density of inclusions in the composite material at the same mass fraction is many times larger if the tubes are individual. This explains why the effective permittivity (and conductivity) of the composite is higher for thinner CNT inclusions at the same their mass fraction. The same is true for MWCNTs. The polarizability of the MWCNT slightly depends on its diameter in quasi-static regime (see Fig. 5 in [25]); the composite with the thinner MWCNTs has higher number density of inclusions and, consequently, higher effective conductivity.

In the dynamic regime, wherein the finite-length effects vanish, the effective conductivity of the composite material is then given by [24].

$$\sigma_{eff}(f) = \frac{2\pi}{3} \sum_m R_m \tilde{n}_m \sigma_{cn}^{(m)}(f), \tag{1.12}$$

where $\sigma_{cn}^{(m)}$ is the axial surface conductivity of SWCNT of type m, and $\tilde{n}_m = \int_0^\infty N_m(L)dL$.

1.3 Spectra of the Effective Parameters of CNT-Based Composites

Figure 1.1a shows the real and imaginary parts of the axial surface conductivity of metallic (12,0) zigzag CNT at the electron relaxation time of 50 fs. One can see from Fig. 1.1a that the real part of the surface conductivity is constant in a wide range 0–0.5THz and equals the dc-conductivity. This is not true for the effective conductivity of the composite comprising finite-length CNTs (see Fig. 1.1b). The spectra for these composites have resonant-like behavior relating to the LPR. If $f_p < \nu$, the value Re(σ_{eff}) is approximately the same at the peak frequency f_p for all the tube lengths, $L = \{1; 5; 10\}$ μm.

Starting from the frequency f_p, the value Re(σ_{eff}) decreases as the frequency decreases. The finite-length effect leads to the screening effect resulting in a decrease of the effective conductivity with decreasing frequency. The LPR frequency is 3 THz and 0.18 THz for the lengths of 1 μm and 5 μm, respectively. As shown in Fig. 1.1c, in the range 0.3–1 THz, the dynamic and quasi-static regimes occur for $L = 5$ μm and $L = 1$ μm tubes, respectively. In practice, there is a distribution of CNTs over the length within composite material. Thus, different regimes (quasi-static and dynamic) of the EM interactions may take place simultaneously providing a large variety of the EM parameters of composites.

Though Fig. 1.1b, c was obtained for composites without percolation, qualitatively the same behavior has been obtained for the films comprising nanotubes of different lengths. Figure 1.2 shows the broad band spectra of the real part of the permittivity and conductivity for the films comprising short- (s-CNT, $L = 0.1 \div 0.3$ μm), medium- (m-CNT, $L = 0.5 \div 2$ μm) and long- (l-CNT, $L = 1 \div 30$ μm) length SWCNTs.

CNT film is a densely packed composite, wherein adjacent CNTs are attached with each other forming conductive network (see typical image of SWCNT film in Fig. 1.3). In spite of the electrical contacts between adjacent CNTs in the film, the finite-length effect is strong for short-length CNTs in sub-teraherz range. As a result, the low-frequency conductivity of s-CNT film is thousand times less than that for the l-CNT film in Fig. 1.2a.

Frequency dependence of the Re(σ_{eff}) in the sub-terahertz and microwave ranges follows the power law

$$Re\left(\sigma_{eff}\right) \cong \sigma_0 \omega^s, \tag{1.13}$$

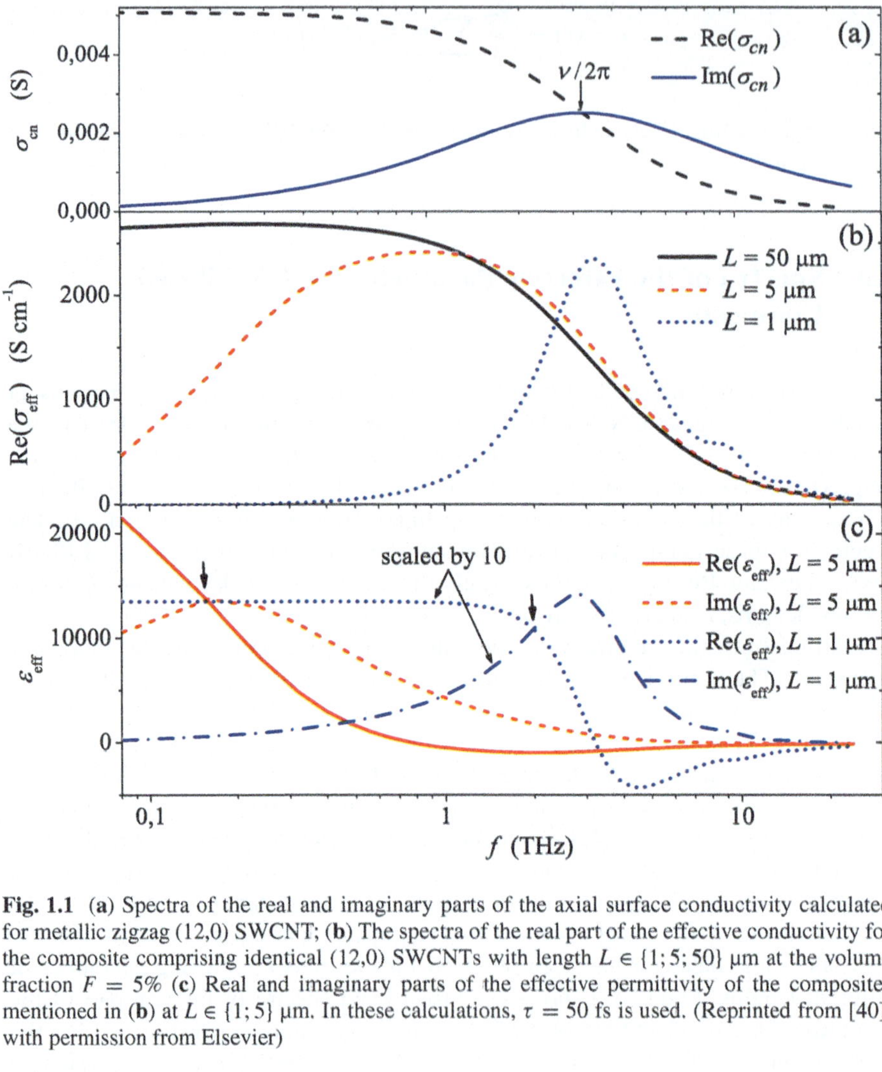

Fig. 1.1 (**a**) Spectra of the real and imaginary parts of the axial surface conductivity calculated for metallic zigzag (12,0) SWCNT; (**b**) The spectra of the real part of the effective conductivity for the composite comprising identical (12,0) SWCNTs with length $L \in \{1; 5; 50\}$ μm at the volume fraction $F = 5\%$ (**c**) Real and imaginary parts of the effective permittivity of the composites mentioned in (**b**) at $L \in \{1; 5\}$ μm. In these calculations, $\tau = 50$ fs is used. (Reprinted from [40], with permission from Elsevier)

with the exponent s being in the range $2 \geq s \geq 0$ in accordance with that was reported for other materials [42]. The value $s = 0$ corresponds to the Drude behavior which occurs for long-length CNTs (see l-CNT in Fig. 1.2a). As soon as the screening effect contributes, the parameter s becomes larger than zero, $s > 0$. The stronger the screening effect, the larger the parameter s is. For example, Fig. 1.4 shows the conductivity spectra for MWCNT thin film and composite comprising 1% wt. of MWCNTs of the same kind but fabricated in different ways. One can see from Fig. 1.4, that the conductivity follows the power law with the exponents $s = 0.21$ for CNT film, $s = 0.62$ and $s = 1.04$ for MWCNTs in epoxy resin and phenol formaldehyde resin, respectively. The difference can be explained as follows. Good

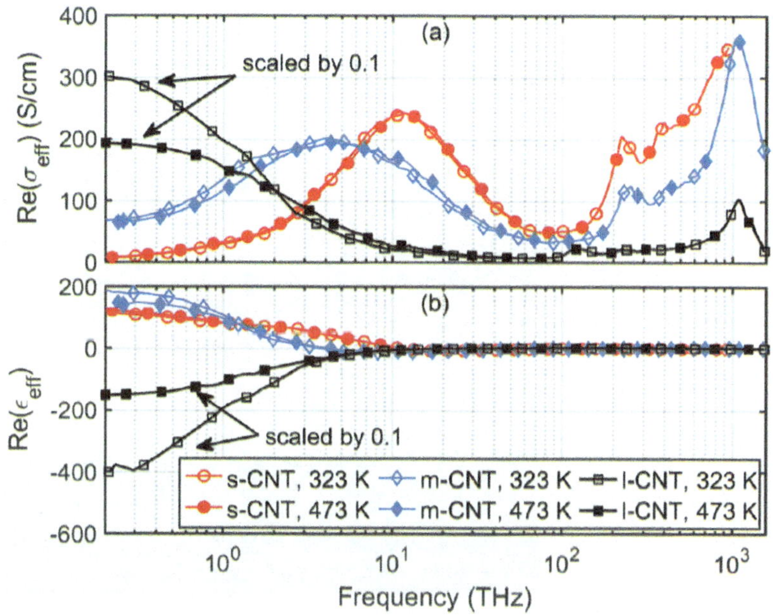

Fig. 1.2 Frequency dependence of (**a**) the real part of the effective conductivity Re(σ_{eff}), and (**b**) the real part of the effective permittivity Re(ε_{eff}), obtained at 323 K and 473 K for s-, m-, and l-CNT films. The lengths of the s-, m-, and l-CNTs are in the ranges 0.1–0.3 μm, 0.5–2 μm, 1–30 μm, respectively. The data for the l-CNTs have been scaled by 0.1. (Reprinted from [41], with permission from IOP Publishing)

contacts between adjacent CNTs cause a small value of s for CNT film. In the 1% wt. composite material, tubes are mostly aggregated, and the screening effect is quite strong. It is more pronounced for CNTs in phenol formaldehyde resin, as tubes become shorter during the preparation process in comparison with that for CNT in the epoxy resin.

Figure 1.5 shows the frequency dependence of the effective permittivity for the MWCNT film. The real and imaginary parts follow power law. We determine the dielectric loss tangent as a ratio

$$tan\delta = Im\left(\varepsilon_{eff} - \varepsilon_h\right)/Re\left(\varepsilon_{eff} - \varepsilon_h\right). \tag{1.14}$$

It does not depend on the permittivity ε_h of dielectric matrix. If the screening effect is week, the value of tanδ is high (see data for CNT film in Fig. 1.6). Let us notice that the larger the exponent s in (1.13), the smaller the loss tangent is. This is clearly seen from comparison of Figs. 1.4 and 1.6 obtained for the same CNT materials.

It is clear from (1.4) that for the composite with identical inclusions, the frequency dependence of the effective permittivity is determined by the polarisability of the inclusions. Distribution of the inclusions over the polarisability α can significantly modify the frequency dependence of the effective permittivity.

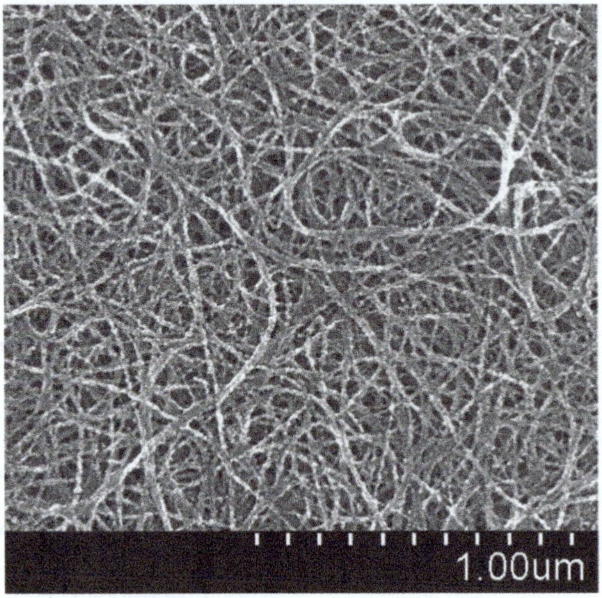

Fig. 1.3 Scanning electron microscopy image of the SWCNT film. SWCNTs are in bundled form

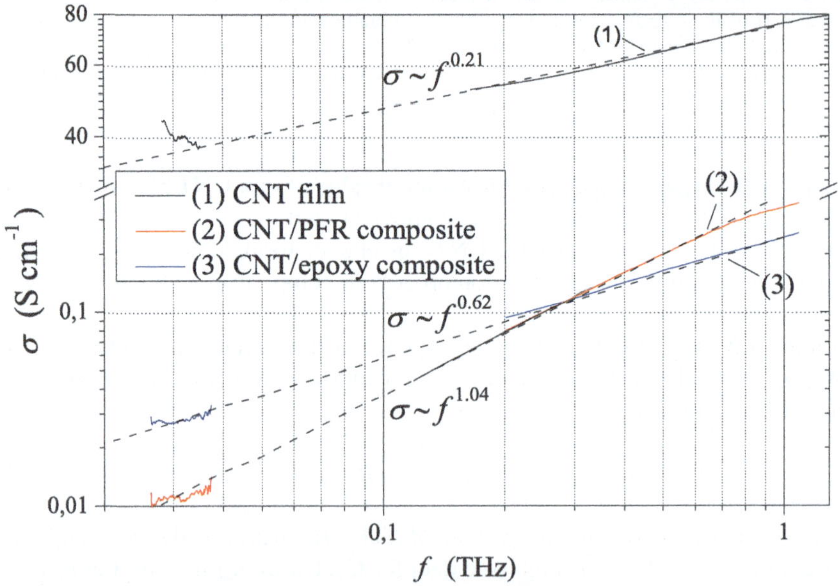

Fig. 1.4 Frequency dependence of the effective conductivity for MWCNT film (1), and composites containing 1% wt. of CNTs in epoxy resin (2), and phenol formaldehyde (PRF) resin (3). (Reprinted from [43], with permission from IEEE)

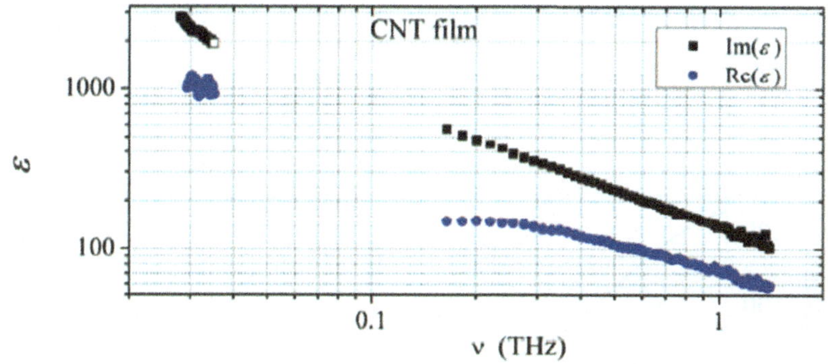

Fig. 1.5 Frequency dependence of the effective permittivity of MWCNT film. (Reprinted from [43], with permission from IEEE)

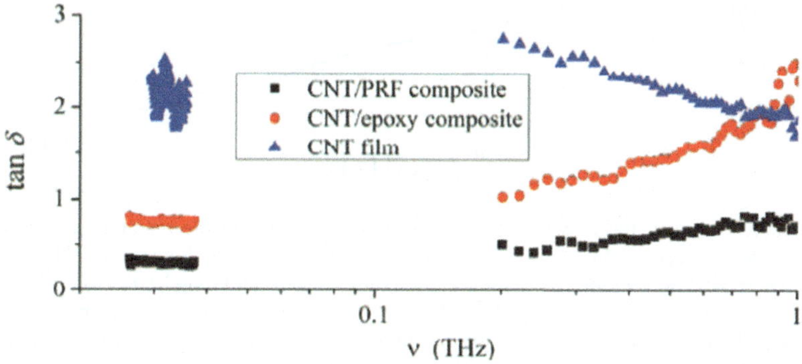

Fig. 1.6 Frequency dependence of the loss tangent tanδ for the CNT film and composites containing 1% wt. of the CNTs in the epoxy and phenol formaldehyde (PRF) resins. (Reprinted from [43], with permission from IEEE)

Figure 1.7 shows the effective permittivity of composite material comprising identical (a) short-, (b) long-length MWCNTs and (c) their mixture. Fig. 1.7a, b demonstrates quasi-static (tan$\delta < < 1$, ∂ Im $(\varepsilon_{eff})/\partial f > 0$) and dynamic (tan$\delta > 1$, ∂ Im $(\varepsilon_{eff})/\partial f < 0$) regimes of the interaction in the range 5–20 GHz, whereas Fig. 1.7c shows the regime where short tubes are in the quasi-static regime and long-tubes are in the dynamic regime of interaction.

As shown in [25], a big variety of the microwave parameters of the CNT-based composite can be divided into three groups:

1. ∂ Im $(\varepsilon_{eff})/\partial f < 0$, Re$(\varepsilon_{eff}) \approx$ *const*, tan$\delta < < 1$, that corresponds predominantly to quasi-static regime of interaction. This regime could be if (i) tubes are short, (ii) adjacent tubes has bad contact with each other, (iii) tubes are mostly agglomerated.

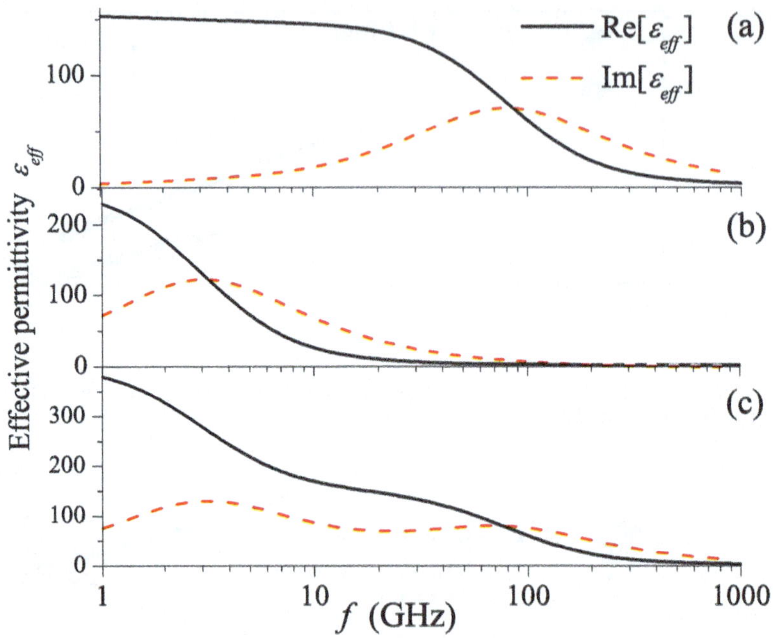

Fig. 1.7 The spectra of the effective relative permittivity of composite material comprising (**a**) short MWCNTs of length $L_1 = 5$ μm at volume fraction 7.5%; (**b**) long MWCNTs of length $L_1 = 30$ μm at volume fraction 0.5%; and (**c**) a mixture of short and long MWCNTs with volume fraction 7.5% and 0.5%, respectively. The MWCNT diameter is 25 nm. Spectra were computed using (1.4). The figure was reprinted with permission [25]. (Copyright (2018) by the American Physical Society)

2. ∂ Im $(\varepsilon_{eff})/\partial f < 0$, ∂ Re $(\varepsilon_{eff})/\partial f < 0$, $\tan\delta > 1$, that corresponds predominantly to the dynamic regime of the interaction. This regime could be if (i) tubes are long, and (or) (ii) adjacent tubes have a good contact with each other forming a homogeneous conductive network.
3. ∂ Im $(\varepsilon_{eff})/\partial f < 0$, ∂ Re $(\varepsilon_{eff})/\partial f < 0$, $\tan\delta \leq 1$, the loss tangent $\tan\delta$ practically does not depend on frequency. This is a result of two types of interactions taking place (i) in the composite with a broad distribution of CNT over the length, and (ii) at non-homogeneous dispersion of the inclusions. Short tubes and tubes electrically isolated from adjacent tubes interact with the EM field in quasi-static regime. Long tubes and tubes actively involved in the formation of the conductive network interact with the EM field in the dynamic regime.

It is suggested in [40], that the screening effect is the main mechanism determining the frequency dependence Re$[\sigma_{eff}(f)]$. The effective surface conductivity of CNT can be calculated as follows

$$\sigma_{cn}^{(eff)} = \frac{\sigma_{cn}}{L} \int_0^L \frac{E_z^{tot}(z)}{E_z^0(z)} dz. \tag{1.15}$$

The value of $E_z^{tot}(z)/E_z^0(z)$ depends on the CNT length and frequency. It also depends on the contact resistance between the adjacent CNTs within a network. Effective conductivity of the composite material is a sum of the effective conductances of the all CNTs in unit volume. As shown in Fig. 1.1b, the value of $\mathrm{Re}[\sigma_{eff}]$ is maximal at the LPR frequency f_p, and $\mathrm{Re}[\sigma_{eff}(f_p)]$ is practically the same for different CNT lengths at the same volume fraction of inclusions; this is true if $f_p < \nu/2\pi$ [40]. The last condition is fulfilled for long-length SWCNTs (for $L > 1$ μm we get $f_p < 3$ THz, whereas $\nu/2\pi > 2$ THz [41]). Then, one can conclude that the screening effect is weak at the frequency f_p and the effective conductivity $\mathrm{Re}[\sigma_{eff}(f_p)]$ is determined by formula (1.12). The value $\mathrm{Re}[\sigma_{eff}(f_p)]$ does not depend on how the inclusions are distributed in the matrix, i.e. it does not depend on the method of the composite preparations. The effectiveness of the inclusion interaction with the EM field at the frequency f_p is maximal and can be estimated as 100%. At lower frequency, the value $\mathrm{Re}[\sigma_{eff}(f)]$ is affected by the screening effect within the inclusions or their aggregates, and it strongly depends on the distribution of the inclusions in the matrix and on the agglomeration effects. The average contribution of the inclusions into the value $\mathrm{Re}[\sigma_{eff}(f)]$ in comparison with maximal possible $\mathrm{Re}[\sigma_{eff}(f_p)]$ can be estimated using the following ratio [40]:

$$g(f, f_0) \equiv \frac{\mathrm{Re}\left[\sigma_{eff}(f)\right]}{\mathrm{Re}\left[\sigma_{eff}(f_0)\right]} \tag{1.16}$$

f_0 is a frequency at which the screening effect vanish and the conductivity follows formula (1.12), it can be assumed $f_0 \approx f_p$. Often, using time-domain spectroscopy, the conductivity of the composite can be obtained in the range 0.2–1 THz. If tube length is high enough ($L > 1$ μm), and consequently loss tangent at 1 THz is more than unity, $\tan\delta \geq 1$, then one can assume the screening effect to be small at 1 THz and one can accept in (1.16) that $f_0 = 1$ THz.

As an example, for the MWCNT film presented in Fig. 1.4, we estimated $g(30\mathrm{GHz}, 1\mathrm{THz}) = 0.49$. Roughly speaking that means that at frequency 30 GHz only 49% of electrons are exposed to the external field, other 61% of electrons are completely screened. For composites presented in Fig. 1.4, for MWCNTs in epoxy resin and phenol formaldehyde resin, we obtained $g(30\mathrm{GHz}, 1\mathrm{THz}) = 0.11$ and $g(30\mathrm{GHz}, 1\mathrm{THz}) = 0.03$, respectively. The difference originates from the preparation process: fabrication method for phenol formaldehyde resin composite leads to CNT cutting. Let us notice that the parameter $g(f)$ grows with frequency.

CNT film could be considered as a perfect composite (see Fig. 1.3). Within the film, CNTs form densely packed conductive network, where the screening effect is minimized. It has maximal possible parameter $g(f, f_0)$, which we define as $g_{film}(f, f_0)$. To compare CNT composite and CNT film, another parameter is proposed [40].

$$g_0(f, f_0) = \frac{g(f, f_0)}{g_{film}(f, f_0)} \tag{1.17}$$

Fig. 1.8 Scanning electron microscopy image of hybrid film at SWCNT density of 11 mg/cm^3. Network of thin SWCNTs is distributed between thick WS_2 nanotubes

Parameters g and g_0 depends on the volume fraction of the inclusions. Recently, thin hybrid films comprising a mixture of conductive SWCNTs and nonconductive WS_2 nanotubes have been fabricated (see Fig. 1.8).

Due to the homogeneous distribution of thin SWCNTs between thick WS_2 nanotubes, the screening effect is not so strong resulting in a weak frequency dependence of the effective conductivity (see Fig. 1.9a). Parameters g and g_0 for hybrid films decrease as volume fraction of SWCNT decreases (see Fig. 1.9b). However, they are still high at low volume fraction: $g_0(30\text{GHz}, 1\text{THz}) > 0.75$ at $\rho/\rho_0 > 0.01$.

1.4 Concentration Dependence of the Conductivity

The concentration dependence of the conductivity of CNT-based composite follows power law

$$\mathrm{Re}\left(\sigma_{eff}\right) = \sigma_0(\varphi - \varphi_0)^{\beta} \tag{1.18}$$

The exponent β in static regime ($f = 0$) is predominantly in the range 1.3–4.0 peaking around $\beta = 2$ [44]. The value of β decreases with increasing frequency reaching unity in the terahertz range. For example, dependencies of the effective conductivity on the volume fraction of SWCNTs for hybrid SWCNT-WS_2 nanotube films (see Fig. 1.8) at 0 Hz, 30GHz, 0.2THz and 1THz is shown in Fig. 1.10 by symbols; lines in Fig. 1.10 are fitted function (1.18) with exponents β= 1.6, 1.3, 1.2, 1.1, respectively [41].

The dependence $\beta(\omega)$ originates from the intertube tunneling of electrons. At low frequencies, the effective conductivity strongly depends not only on the concentration of the inclusions, but also on the number of contacts between adjacent

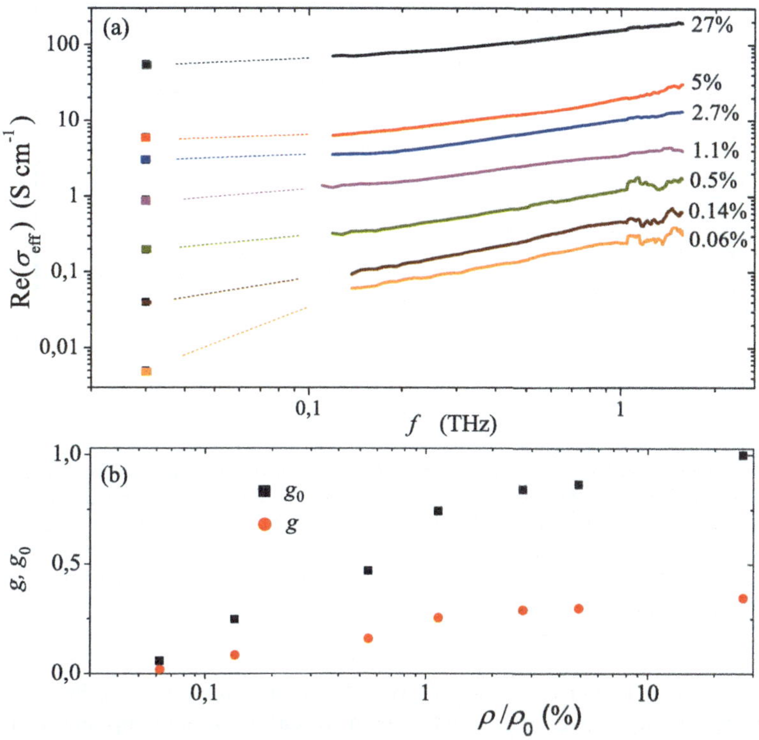

Fig. 1.9 (**a**) Frequency dependence of Re(σ_{eff}) for the film comprising SWCNTs and WS_2 nanotubes at different CNT volume fractions $\rho/\rho_0 \in (0.06; 0.14; 0.5; 1.1; 2.7; 5)\%$; ρ_0 is a density of CNTs; $\rho_0 = 2.2$ g/cm^3 is the density of graphite; for pure CNT film $\rho/\rho_0 = 27$ %; (**b**) The parameters $g(30\text{GHz}, 1\text{THz})$ and $g_0(30\text{GHz}, 1\text{THz})$ obtained from data presented in (**a**) versus ρ/ρ_0. (Reprinted from [40], with permission from Elsevier)

tubes. The intertube contacts influence the charge distribution on CNT surface and, consequently, the total field along the tubes.

In the terahertz range, the influence of contacts is weak. The current is excited even in individual CNTs isolated from others. For sufficiently long individual tubes (>1 μm) the screening effect is small. That is why β is close to unity. Almost linear dependence of the effective conductivity on the tube fraction in the terahertz range can be used to estimate the concentration of the inclusions in the composite materials.

In the microwave range, the influence of the contacts is strong. Density dependence for SWCNT and MWCNT sponges in the microwave range was measured in the range 26–37 GHz by means of the compression of the same sample [45] (see data in Fig. 1.11). It follows power law with exponents 1.7 ± 0.2 and 2 ± 0.2 for MWCNTs and SWCNTs, respectively. Moreover, the value β can be slightly decreased by the addition of the nonconductive component, which decreases the

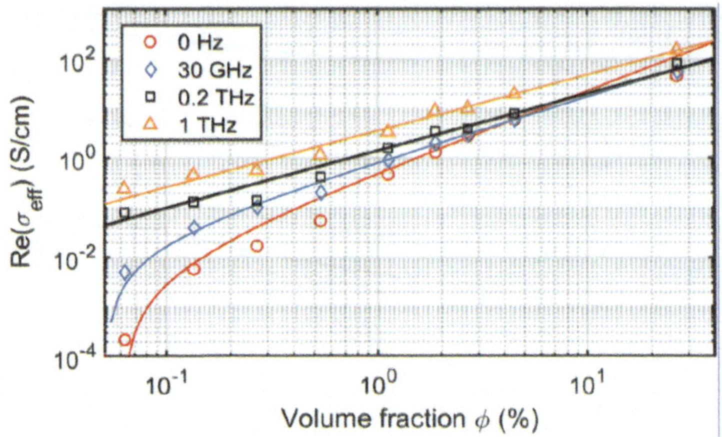

Fig. 1.10 Effective conductivity Re(σ_{eff}) of SWCNT-composites versus CNT volume fraction φ for frequencies 0 Hz, 30 GHz, 0.3 THz and 1 THz. Points are experimental data and lines are fitted power functions (1.18) with β= 1.6, 1.3, 1.2, 1.1 and $\varphi_0 = 0.06\%, 0.05\%, 0\%, 0\%$ for 0 Hz, 30 GHz, 0.3 THz and 1 THz, respectively. Volume fraction was taken as ρ/ρ_0 is a density of CNTs; $\rho_0 = 2.2$ g/cm^3 is the density of graphite; for pure CNT film $\rho/\rho_0 = 0.27$. (Reprinted from [41], with permission from IOP Publishing)

number of contacts between inclusions [45]. Thus, the EM response of CNT-sponges in the microwave range can be tuned smoothly through applied mechanical stress to them.

1.5 Conclusions

Three regimes of the interaction of CNTs with the EM field occur in the CNT-based composite media. They are quasi-static, resonant and dynamic. The main mechanism hindering the EM interaction is the screening effect which takes place in the quasi-static regime. Frequency dependence of the effective conductivity of the CNT-based composite follows power law. It can be used to estimate how effectively the inclusions interact with the EM field. CNT thin film is proposed to use as a perfect composite with maximal possible EM interaction. Concentration dependence of the effective permittivity follows power law with the exponent in the range 1÷2.2. The exponent indicates the influence of the contact resistance on the effective conductivity of the composite. In the terahertz range, the exponent is close to unity that can be used to estimate the concentration of the inclusions by means of the terahertz spectroscopy.

Acknowledgements This research was partially supported by the Belarusian Republican Foundation for Fundamental Research (BRFFR) under project F18Kor-002 by the H2020-MSCA-RISE-2014 project 644076 CoExAN, and the MSCA-RISE-2016 project 734164 "Graphene-3D".

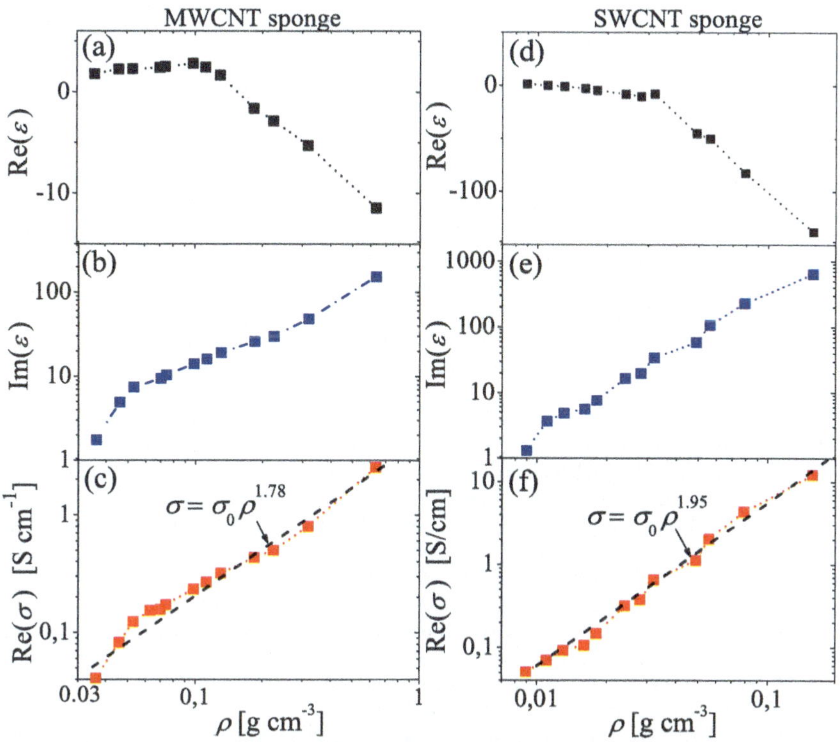

Fig. 1.11 Effective permittivity, ε_{eff}, and conductivity, Re(σ_{eff}), at 32 GHz versus the density of (**a**, **b**, **c**) MWCNT and (**d**, **e**, **f**) SWCNT sponges. (Reprinted from [45], with permission from IOP Publishing)

References

1. Saito R, Dresselhaus G, Dresselhaus MS (2004) Physical properties of carbon nanotubes. Imperial College Press, Singapore
2. Reich S, Thomsen C, Maultzsch J (2004) Carbon nanotubes. Basic concepts and physical properties. Wiley-VCH, Berlin
3. Tasaki S, Maekawa K, Yamabe T (1998) π-Band contribution to the optical properties of carbon nanotubes: Effects of chirality. Phys Rev B 57:9301
4. Slepyan GY, Maksimenko SA, Lakhtakia A, Yevtushenko O, Gusakov AV (1999) Electrodynamics of carbon nanotubes: Dynamic conductivity, impedance boundary conditions, and surface wave propagation. Phys Rev B 60:17136–17149
5. Maksimenko SA, Slepyan GY (2000) Electrodynamic properties of carbon nanotubes. In: Singh ON, Lakhtakia A (eds) Electromagnetic fields in unconventional materials and structures. Wiley, New York, pp 217–255
6. Wang F, Itkis ME, Haddon RC (2010) Enhanced electromodulation of infrared transmittance in semitransparent films of large diameter semiconducting single-walled carbon nanotubes. Nano Lett 10:937
7. Blackburn JL, Barnes TM, Beard MC, Kim YH, Tenent RC, McDonald TJ, To B, Coutts TJ, Heben MJ (2008) Transparent conductive single-walled carbon nanotube networks with precisely tunable ratios of semiconducting and metallic nanotubes. ACS Nano 2:1266–1274

8. Zhang Q, Háróz EH, Jin Z, Ren L, Wang X, Arvidson RS, Lüttge A, Kono J (2013) Plasmonic nature of the terahertz conductivity peak in single-wall carbon nanotubes. Nano Lett 13:5991–5996
9. Nemilentsau AM, Shuba MV, Slepyan GY, Kuzhir PP, Maksimenko SA, D'yachkov PN, Lakhtakia A (2010) Substitutional doping of carbon nanotubes to control their electromagnetic characteristics. Phys Rev B 82:235424
10. Hanson GW (2005) Fundamental transmitting properties of carbon nanotube antennas. IEEE Trans Antennas Prop 53:3426–3435
11. Slepyan GY, Shuba MV, Maksimenko SA, Lakhtakia A (2006) Theory of optical scattering by achiral carbon nanotubes and their potential as optical nanoantennas. Phys Rev B 73:195416
12. Burke PJ, Li S, Yu Z (2006) Quantitative theory of nanowire and nanotube antenna performance. IEEE Trans Nanotechnol 5:314–334
13. Shuba MV, Slepyan GY, Maksimenko SA, Thomsen C, Lakhtakia A (2009) Theory of multiwall carbon nanotubes as waveguides and antennas in the infrared and the visible regimes. Phys Rev B 79:155403
14. Shuba MV, Maksimenko SA, Lakhtakia A (2007) Electromagnetic wave propagation in an almost circular bundle of closely packed, metallic, carbon nanotubes. Phys Rev B 76:155407
15. Shuba MV, Slepyan GY, Maksimenko SA, Hanson GW (2010) Radiofrequency field absorption by carbon nanotubes embedded in a conductive host. J Appl Phys 108:114302
16. Hassan AM, Vargas-Lara F, Douglas JF, Garboczi EJ (2016) Electromagnetic resonances of individual single-walled carbon nanotubes with realistic shapes: a characteristic modes approach. IEEE Trans Antenn Propag 64:2743–2757
17. Shuba MV, Melnikov AV, Kuzhir PP, Maksimenko SA, Slepyan GY, Boag A, Mosca Conte A, Pulci O, Bellucci S (2017) Integral equation technique for scatterers with mesoscopic insertions: application to a carbon nanotube. Phys Rev B 96:205414
18. Hao J, Hanson GW (2016) Electromagnetic scattering from finite-length metallic carbon nanotubes in the lower IR bands. Phys Rev B 74:035119
19. Nefedov IS (2010) Electromagnetic waves propagating in a periodic array of parallel metallic carbon nanotubes. Phys Rev B 82:155423
20. Mikki S, Kishk A (2018) An efficient algorithm for the analysis and design of carbon nanotube photonic crystals. Prog Electromagnet Res C 83:83–96
21. Gong S, Zhu ZH, Haddad EI (2013) Modeling electrical conductivity of nanocomposites by considering carbon nanotube deformation at nanotube junctions. J Appl Phys 114:074303
22. Bao WS, Meguid SA, Zhu ZH, Pan Y, Weng GJ (2013) Effect of carbon nanotube geometry upon tunneling assisted electrical network in nanocomposites. J Appl Phys 113:234313
23. Waterman PC, Truell R (1961) Multiple scattering of waves. J Math Phys 2:512
24. Slepyan GY, Shuba MV, Maksimenko SA, Thomsen C, Lakhtakia A (2010) Terahertz conductivity peak in composite materials containing carbon nanotubes: theory and interpretation of experiment. Phys Rev B 81:205423
25. Shuba MV, Melnikov AV, Paddubskaya AV, Kuzhir PP, Maksimenko SA, Thomsen C (2013) The role of finite size effects in the microwave and sub-terahertz electromagnetic response of multiwall carbon nanotube based composite: theory and interpretation of experiment. Phys Rev B 88:045436
26. Burke P (2002) Luttinger liquid theory as a model of the gigahertz electrical properties of carbon nanotubes. IEEE Trans Nanotechnol 1:129–144
27. Shuba MV, Paddubskaya AG, Plyushch AO, Kuzhir PP, Slepyan GY, Maksimenko SA, Ksenevich VK, Buka P, Seliuta D, Kasalynas I, Macutkevic J, Valusis G, Thomsen C, Lakhtakia A (2012) Experimental evidence of localized plasmon resonance in composite materials containing single-wall carbon nanotubes. Phys Rev B 85:165435
28. Nemilentsau AM, Slepyan GY, Maksimenko SA (2007) Thermal radiation from carbon nanotubes in the terahertz range. Phys Rev Lett 99:147403
29. Shuba MV, Maksimenko SA, Slepyan GY (2009) Absorption cross-section and near-field enhancement in finite-length carbon nanotubes in the terahertz-to-optical range. J Comput Theor Nanosci 6:2016–2023

30. Shuba MV, Paddubskaya AG, Kuzhir PP, Maksimenko SA, Valusis G, Ivanov M, Banys J, Ksenevich V, Hanson GW (2017) Observation of the microwave near-field enhancement effect in suspensions comprising single-walled carbon nanotubes. Mat Res Exp 4:075033
31. Shuba MV, Paddubskaya A, Kuzhir PP, Maksimenko SM, Flahaut E, Fierro V, Celzard A, Valusis G (2017) Short-length carbon nanotubes as building blocks for high dielectric constant materials in the terahertz range. J Phys D 50:08LT01
32. Shuba MV, Maksimenko SA (2016) Carbon nanotube based composites as materials for terahertz application. J Appl Spectrosc 83:753–754
33. Ozbay E (2006) Plasmonics: merging photonics and electronics at nanoscale dimensions. Science 311:189–193
34. Ju L, Geng B, Horng J, Girit C, Martin M, Hao Z, Bechtel HA, Liang X, Zettl A, Shen YR, Wang F (2011) Graphene plasmonics for tunable terahertz metamaterials. Nat Nanotech 6:630–634
35. Shuba MV, Paddubskaya AG, Kuzhir PP, Slepyan GY, Seliuta D, Kasalynas I, Valusis G, Lakhtakia A (2012) Effects of inclusion dimensions and p-type doping in the terahertz spectra of composite materials containing bundles of single-wall carbon nanotubes. J Nanophoton 6:061707
36. Shuba MV, Paddubskaya AG, Kuzhir PP, Maksimenko SA, Ksenevich V, Niaura G, Seliuta D, Kašalynas I, Valusis G (2012) Soft cutting of single-wall carbon nanotubes by low temperature ultrasonication in a mixture of sulfuric and nitric acids. Nanotechnology 23:495714
37. Morimoto T, Joung S-K, Saito T, Futaba DN, Hata K, Okazaki T (2014) Length-dependent plasmon resonance in single-walled carbon nanotubes. ACS Nano 8:9897–9904
38. Morimoto T, Okazaki T (2015) Optical resonance in far-infrared spectra of multiwalled carbon nanotubes. Appl Phys Express 8:055101
39. Shuba MV, Yuko DI, Kuzhir PP, Maksimenko SA, Chigir GG, Pyatlitski AN, Sedelnikova O, Okotrub AV, Lambin P (2018) Localized plasmon resonance in boron doped multi-walled carbon nanotubes. Phys Rev B 97:205427
40. Shuba MV, Yuko DI, Kuzhir PP, Maksimenko SA, Kanygin MA, Okotrub AV, Tenne R, Lambin P (2018) How effectively do carbon nanotube inclusions contribute to the electromagnetic performance of a composite material? Estimation criteria from microwave and terahertz measurements. Carbon 129:688–694
41. Karlsen P, Shuba MV, Beckerleg C, Yuko DI, Kuzhir PP, Maksimenko SA, Ksenevich V, Viet H, Nasibulin AG, Tenne R, Hendry E (2018) Influence of nanotube length and density on the plasmonic terahertz response of single-walled carbon nanotubes. J Phys D Appl Phys 51:014003
42. Jonscher AK (1999) Dielectric relaxation in solids. J Phys D Appl Phys 32:R57
43. Shuba M, Yuko D, Bychanok D, Liubimau A, Meisak D, Bochkov I, Kuzhir P (2017) Comparison of the electrical conductivity of polymer composites in the microwave and terahertz frequency ranges, IEEE International Conference on Microwaves, Antennas, Communications and Electronic Systems (COMCAS): 1–3
44. Bauhofer W, Kovacs JZ (2009) A review and analysis of electrical percolation in carbon nanotube polymer composites. Comput Sci Technol 69:1486–1498
45. Shuba MV, Yuko DI, Kuzhir PP, Maksimenko SA, De Crescenzi M, Scarselli M (2018) Carbon nanotube sponges as tunable materials for electromagnetic applications. Nanotechnology 29:375202

Chapter 2
Electrophysical Properties of $Sr_2FeMoO_{6-\delta}$ Ceramics with Dielectric Shells

Nikolay Kalanda, Marta Yarmolich, Sergey Demyanov, Alexander Petrov, Vasil M. Garamus, Herman Terryn, Jon Ustarroz, and Nikolai A. Sobolev

Abstract Magnetic metal-oxide compounds with high magnetoresistance (MR) values have attracted an extraordinary interest of researchers and engineers due to their possible spintronic applications. Among these materials, the strontium ferromolybdate $Sr_2FeMoO_{6-\delta}$ (SFMO) has been relatively less known compared to cobaltites and manganites despite its 100% spin polarization of the conduction electrons and a high Curie temperature. In the present chapter, a stable fabrication technology and a systematic analysis of nanocomposites consisting of SFMO grains with $SrMoO_4$ (SMO) dielectric shells are presented. SFMO-SMO nanocomposites were fabricated as follows: synthesis of the SFMO single-phase nanopowders by a modified citrate-gel technique; compaction under high pressure; thermal treatment for the formation of dielectric shells on the grain surface. The nanocomposite exhibits a transitional behavior of the conductivity type from metallic, which is characteristic of the SFMO, to semiconducting in the temperature range from 4 to 300 K in applied magnetic fields up to 10 T. A negative MR is observed due to the tunneling of spin-polarized charge carriers through dielectric interlayers. The MR value reaches 43% in a field of 8 T at 10 K. The dielectric shell thickness was determined to be about 10 nm from the electrical breakdown voltage on the current-voltage characteristics. The observed electrical breakdown is found to be a reversible process determined by the impact ionization of atoms in the dielectric in strong electric fields depending on the electrons knocked-out from $SrMoO_4$. It

N. Kalanda (✉) · M. Yarmolich · S. Demyanov · A. Petrov
Scientific-Practical Materials Research Centre, NAS of Belarus, Minsk, Belarus
e-mail: kalanda@physics.by

V. M. Garamus
Helmholtz-Zentrum Geesthacht: Centre for Materials and Coastal Research, Geesthacht, Germany

H. Terryn · J. Ustarroz
Vrije Universiteit Brussel, Brussels, Belgium

N. A. Sobolev
Departamento de Física and I3N, Universidade de Aveiro, Aveiro, Portugal

A. Maffucci, S. A. Maksimenko (eds.), *Fundamental and Applied Nano-Electromagnetics II*, NATO Science for Peace and Security Series B: Physics and Biophysics, https://doi.org/10.1007/978-94-024-1687-9_2

was found that the MR changes the sign from negative to positive in the electrical breakdown region, revealing giant MR properties.

Keywords Strontium ferromolybdate · Ferrimagnetic · Sol-gel synthesis · Dielectric shells · Electrical resistivity · Tunneling magnetoresistance

2.1 Introduction

The strontium ferromolybdate $Sr_2FeMoO_{6-\delta}$ (SFMO) compound with the double perovskite structure has recently attracted much interest as a prospective spintronic material exhibiting a high Curie temperature ($\sim$ 420 K) and a high degree of spin polarization reaching $\sim$ 100% [1–7]. The SFMO belongs to the class of half-metals, with the magnetic ordering persisting well above room temperature. According to the band structure of the SFMO, the spin-up electrons of the $Fe^{+}(3d^5)$ cations are placed in the valence band below the Fermi level on the $Fe(t_{2g})$ and $Fe(e_g)$ orbitals, and the spin-down electrons of the $Mo^{5+}(4d^1)$ cations sit in the conduction band on the $Mo(t_{2g})$ and $Fe(t_{2g})$ electron orbitals. As a result of the neutron powder diffraction investigations, it has been found that an energy gap of 0.8 eV is formed between the $Fe(e_g)$ upper electron levels and the Fermi level. This enables the built-up of a 100% polarization of the free electrons. The presence of spin-down electrons on the $Mo(t_{2g})$ orbitals leads to the appearance of metallic conductivity and promotes a strong coupling in the Fe – O – Mo – O – Fe chain by means of the double-exchange mechanism.

Kobayashi et al. first reported a negative magnetoresistance (MR) in SFMO in low magnetic fields [8], where it became clear that the degree of superstructural ordering (*P*) of iron and molybdenum cations influences the MR of SFMO. The *P* value is determined by the density of point antisite defects of the $[Fe]_{Mo}$, $[Mo]_{Fe}$ type, which depends on the synthesis conditions [9–11]. A considerable role on the MR value as well as on other electrical properties is played by the state of boundaries between adjacent SFMO grains [12, 13]. Yuan et al. [14] were the first to observe that a decrease of the particle size leads to an enhancement of the MR effect in SFMO nanopowders. In this regard, wet chemistry methods (such as the citrate-gel technique) are considered to be most effective for the SFMO nanopowder synthesis, since the nano-size of the grains implies an increase of the contribution of the intergrain boundaries in the charge transfer processes [15–17].

It has been known that the energy band structure of the double perovskites is favorable for the realization of the tunneling magnetoresistance (TMR). Submicron grains with a narrow size distribution are required for the achievement of large TMR values. A certain attention has been payed to heterostructures on the base of double perovskites [13, 18–23], among which composite structures such as $Sr_2Fe_{1-x}Al_xMoO_6$, $Sr_2Fe_{1-x}Ag_xMoO_6$ are distinguished [20–24]. For further understanding of the electronic and magnetic properties of SFMO nanocomposites, a stable and reproducible fabrication process is necessary. Unfortunately, the manufacture of these structures is associated with a number of important practical problems, such as a very complicated process technology and the need for an expensive equipment. Therefore, the development of novel and simple creation methods of SFMO-based nanocomposites, allowing to keep a large MR and a high Curie temperature, still remains a challenge.

2.2 Experimental Procedures

The citrate-gel technique was used for the synthesis of the SFMO nanopowders. This is a method for the synthesis of highly homogeneous and highly dispersed oxide materials using complexation with intermediate gel production [25]. It has been found that the obtaining of the single-phase $Sr_2FeMoO_{6-\delta}$ nanopowders in this way is problematic due to the presence of secondary phases [15]. As the particle agglomeration was observed in the SFMO nanopowders after the synthesis, the ultrasonic dispersion was used. This technique made it possible to obtain highly dispersive homogeneous particles in a liquid medium. Ethyl alcohol was used in the process, as it does not chemically interact with the material, to avoid the compound decomposition. The non-ionic surface active Tween 20 compound was added to the suspension to prevent the agglomeration. The ultrasound dispersion has been implemented by the Bandelin HD2200 setup. Measurements of the SFMO individual grain size were carried out by means of the dynamic light scattering (DLS) technique using the Zetasizer nanoparticles analyzer.

The powders were compacted into pellets with a diameter of 10 mm at pressures of 1–4 GPa, in the temperature range from 300 to 1100 K. The samples in the form of parallelepipeds with dimensions of $8 \times 3 \times 3$ mm^3 were cut from the pellets. Indium contacts were deposited by the ultrasound soldering, where the current contacts were deposited on the sample face planes and the potential ones, on the long-axis surface.

The phase content and superstructural ordering degree ($P = (2 \cdot SOF - 1) \cdot 100\%$, where SOF is the Seat Occupancy Factor) of the powders were determined by the X-ray diffraction (XRD) using the Siemens D5000 setup employing CuK_α radiation. The XRD patterns were refined using the ICSD-PDF2 (Release 2000) database and the FULLPROF and PowderCell Rietveld programs.

The X-ray photoelectron spectroscopy (XPS) measurements were carried out using the PHI 5600 (Physical Electronics) photoelectronic spectrometer with a monochromatic X-ray source Al$K\alpha$ 1486.6 eV at 350 W. SEM measurements were performed using the JEOL JSM-7000F field emission scanning electron microscope.

Electrical resistivity and MR were investigated by means of the Liquid Helium-Free High-Field Measurement System (Cryogenic Ltd., UK), in the temperature range from 4 to 300 K in constant magnetic fields up to 10 T. The measurements were carried out upon commutation of the magnetic field and electrical current in order to avoid the influence of parasitic thermopower.

2.3 Synthesis of the Single-Phase SFMO Samples

Modified citrate-gel technique was used for the SFMO synthesis by correcting the synthesis and thermal annealing modes, which was possible due to investigations of phase transformation degree processes using XRD and energy-dispersive X-ray spectroscopy (EDX) measurements [26]. In the course of the synthesis of SFMO nanopowders by the citrate-gel technique, the following high purity reagents were used: strontium nitrate $Sr(NO_3)_2$, iron nitrate $Fe(NO_3)_3{\cdot}9H_2O$, ammonium molybdate $(NH_4)_6Mo_7O_{24}$, and citric acid monohydrate $C_6H_8O_7{\cdot}H_2O$. To obtain the colloid sol, aqueous solutions of the strontium and iron monohydrates were mixed in the molar ratio of (2Sr)/(Fe). Citric acid was added in the proportion of (6.5 citric acid)/(Fe). After that, the prepared aqueous solution of $(NH_4)_6Mo_7O_{24}$ was added to the solution of iron and molybdenum nitrates in the proportion of (Mo)/(Fe). Then, under mixing in an IKA C-MAG HS7 setup, ethylenediamine was added, until the $pH = 4$ value of the solution was achieved. Subsequently, the solution was dried at 353 K. The obtained solid sediment was placed in a furnace at 373 K with a subsequent heating at a rate of 0.4 K/min up to 473 K. It was cured for 18 h and then cooled down together with the furnace. At this step a solid foam was obtained, which was ground up and heat-treated at 773 K in an oxygen atmosphere under a pressure of 0.21×10^5 Pa for 10 h. The final annealing permitting the synthesis of SFMO was performed in a gas mixture of 5% H_2/Ar with a final heat treatment at 1223 K for 4 h [26, 27].

As a result, the single-phase SFMO compound has been obtained, which was confirmed by the XRD data, as shown in Fig. 2.1a. The XRD results have shown that the degree of superstructural ordering (P) decreases with the increase of the pH value of the initial solution. A decrease of the P value leads to an electronic density redistribution and a passage of a part of the iron and molybdenum cations to lower-spin states: $Fe^{3+} \rightarrow Fe^{2+}$ and $Mo^{5+} \rightarrow Mo^{6+}$ [26, 27]. In our powders, the condition of $pH = 4$ has been found to be optimal, with the P value reaching 88%.

From the analysis of SEM data, the average size of SFMO single-phase powder agglomerates was found to be 0.8–1 μm. After the ultrasound dispersion with optimal parameters in accordance with the technique described above, the smallest average grain size concluded from the DLS data was about 70 nm with a clearly expressed peak, as shown in Fig. 2.1b.

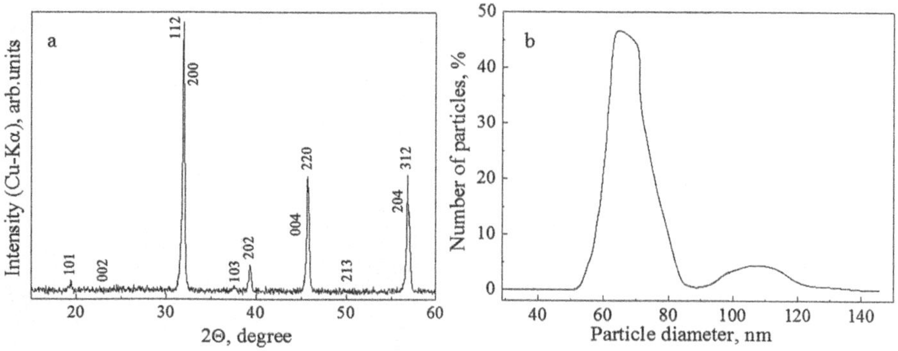

Fig. 2.1 Characterization of the SFMO single-phase powders: (**a**) XRD spectrum; (**b**) DLS analysis after ultrasonic dispersion

A goal of the present investigation is concerned with the magnetoresistive properties of the ferrimagnetic SFMO-dielectric composite. After the acquisition of the SFMO nanosized powders, which are the base of the nanocomposite, it was necessary to solve the problem of the creation of dielectric interlayers between the grains. Moreover, these interlayers should be artificially formed as shells around the grains, closely muffling up each grain.

$Sr_2FeMoO_{6-\delta}$ nanosized powders were pressed into pellets at various pressures and temperatures. Compacting conditions in the range of pressures $p = 1$–4 GPa and temperatures $T = 300$–1100 K have been optimized in accordance with the data of microstructural and X-ray phase analyses, as well as the electrical resistivity, measured in the temperature range from 4.2 to 300 K. It has been found that at $T \leq 500$ K for all the pressure values the samples contained cracks and were mechanically unstable. A temperature increase during the compacting of the SFMO powders up to $T = 800$ K promotes a further improvement of the microstructure and a practically complete absence of microcracks (Fig. 2.2a, b). A further increase of the compacting temperature up to 1100 K leads to a phase separation with the appearance of Fe and $SrMoO_4$ inclusions (Fig. 2.2c).

The optimization of the dense single-phase samples, suitable for the formation of $SrMoO_4$ dielectric shells around $Sr_2FeMoO_{6-\delta}$ grains, has been carried out considering the results of the transport measurements (Fig. 2.3).

An analysis of these figures makes it possible to clearly elucidate the influence of the compaction pressure and temperature on the charge transfer processes in the single-phase SFMO. After compaction at low pressures (Fig. 2.3a) the dependence has a clearly expressed semiconducting character, which indicates the absence of a close contact between granules. An increase of the compaction pressure up to 3 GPa (Fig. 2.3b) leads to a more complicated behaviour of the electrical resistivity:

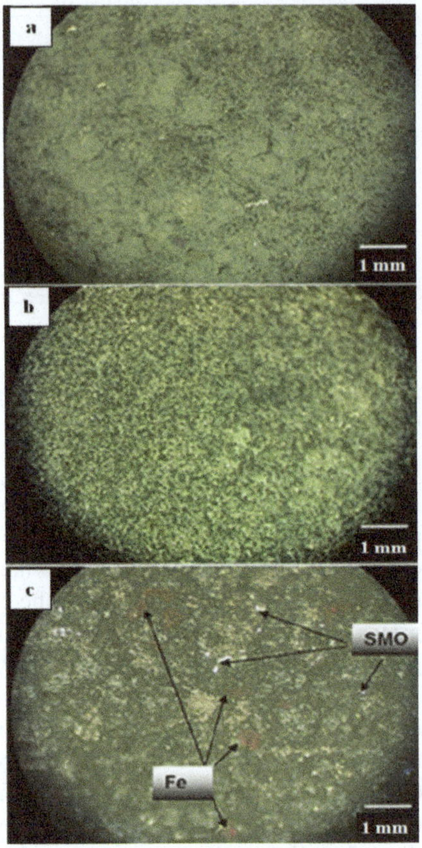

Fig. 2.2 Microstructure of $Sr_2FeMoO_{6-\delta}$ pellets compacted under pressures: (**a**) 1 GPa, (**b**) 4 GPa at 800 K, (**c**) 4 GPa at 1100 K

a semiconductor-type conductivity is observed in the temperature range from 4 to 150 K, and further a metallic conductivity takes place up to 300 K. At low temperatures the conductivity increases by an order of magnitude as compared to the dependence shown in Fig. 2.3a. Still, this means that the electrical transport between the granules is unessential. At temperatures above 150 K the charge carrier scattering on phonons becomes dominant, which is sufficient for the appearance of the metallic conductivity. The optimal pressing conditions (4 GPa) have been confirmed by the temperature dependence of the resistivity in the temperature range from 4 to 300 K, as shown in Fig. 2.3c. One clearly sees the metallic conductivity (which is characteristic of the single-phase SFMO), and consequently there is a high density of the electrical contacts between grains. The dielectric shell formation on the grains' surface has been carried out in these powder samples. Interestingly, the compaction temperature increase up to 1100 K leads to a phase separation with the appearance of Fe and $SrMoO_4$ inclusions (as seen in the microstructure, Fig. 2.2c).

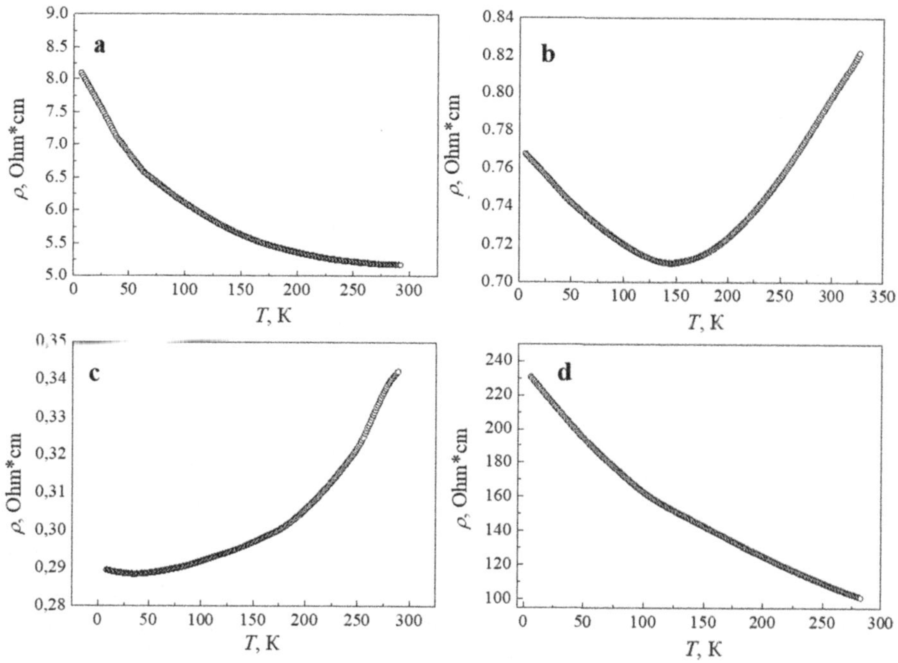

Fig. 2.3 Temperature dependences of the electrical resistivity of the single-phase SFMO at various pressures and temperatures of compaction: (**a**) $p = 1$ GPa at $T = 800$ K; (**b**) $p = 3$ GPa at $T = 800$ K; (**c**) $p = 4$ GPa at $T = 800$ K; (**d**) $p = 4$ GPa at $T = 1100$ K

2.4 Formation and Electrical Transport Characteristics of the SFMO Composite with Dielectric Shells

Additional annealing treatments were performed to modify the SFMO granules surface in the following series of samples: unannealed (SFMO-a), annealed at $T = 700$ K and $p(O_2) = 10$ Pa for $t = 3$ h (SFMO-b), and annealed at $T = 700$ K and $p(O_2) = 10$ Pa for $t = 5$ h (SFMO-c). According to the XRD data, the crystal structure parameters a, b and c, as well as the degree of the superstructural ordering of the SFMO-a, SFMO-b and SFMO-c samples, did not change after annealing at sufficiently low temperatures. Nevertheless, the $SrMoO_4$ phase reflexes were detected in the XRD pattern, and the content of this phase did not exceed 7.5% (Fig. 2.4).

The XPS analysis indicates a change of the valence state of the molybdenum ions on the SFMO grains' surface, caused by the formation of strontium molybdate. Longer annealing in an Ar flow at $T = 700$ K promotes the transition of the surface molybdenum cations to a different valence state, $Mo^{5+} \rightarrow Mo^{6+}$, which indicates the possibility of the $SrMoO_4$ compound formation. The analysis of the chemical state of the molybdenum cations in the SFMO-a,b,c powders by the XPS technique

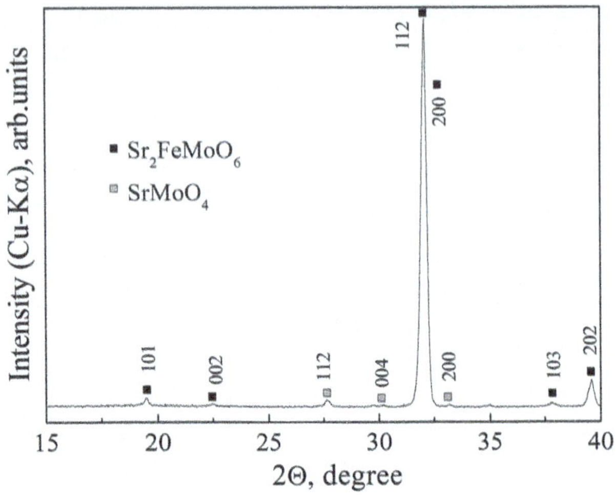

Fig. 2.4 XRD pattern of the SFMO powder with dielectric shells

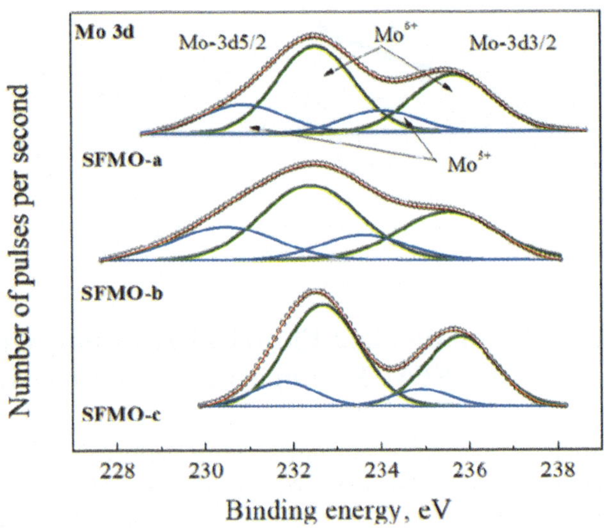

Fig. 2.5 XPS spectra of the Mo_{3d} framing levels of the SFMO-a,b,c samples

has indicated that the Mo_{3d} framing level is the spin-orbital doublet, and it consists of two $Mo_{3d}{}^{5/2}$ and $Mo_{3d}{}^{3/2}$ sub-levels (Fig. 2.5).

From the splitting of the XPS spectra of iron it has been concluded that the spectrum of the Mo_{3d} framing level is a superposition of four components: two for the $Mo_{3d}{}^{5/2}$ sub-level, and two for the $Mo_{3d}{}^{3/2}$ sub-level, indicating the presence of the Mo^{5+} and Mo^{6+} valence states (these areas are highlighted by blue and green colors, correspondingly). In the molybdenum energy spectra the distance between the doublets was 3.13 eV, and the $(Mo_{3d}{}^{5/2})/(Mo_{3d}{}^{3/2})$ peaks ratio equaled 3/2 [28]. Thus, we conclude a binding energy increase for Mo^{6+}, as well as a rise of the Mo^{6+}

Table 2.1 The binding energy E_B of the Mo_{3d} framing levels and the percentage ratio of molybdenum cations valence states in the $Sr_2FeMoO_{6-\delta}$ compound, obtained at different exposure time values in an Ar flow

	E_B ($Mo_{3d}{}^{5/2}$), eV		E_B ($Mo_{3d}{}^{3/2}$), eV	
	Mo^{6+}	Mo^{5+}	Mo^{6+}	Mo^{5+}
SFMO-a	232.51 (68%)	230.86 (32%)	235.68 (68%)	233.98 (32%)
SFMO-b	232.40 (73%)	230.46 (27%)	235.55 (73%)	233.63 (27%)
SFMO-c	232.67 (79%)	231.76 (21%)	235.80 (79%)	234.91 (21%)

concentration with increasing annealing time (t) (see Table 2.1). On the contrary, the Mo^{5+} binding energy and the respective ion concentration decrease with rising t.

Therefore, the XPS data show that the molybdenum cations in the $Sr_2FeMoO_{6-\delta}$ grain surface layers are passing to a new valence state, and the Mo^{6+} concentration rises with annealing time increasing up to 3 h at $T = 700$ K and $p(O_2) = 10$ Pa. After the removal of the $SrMoO_4$ surface layers by means of a reduction at $T = 1170$ K in a continuous flow of the 5%H_2/Ar gas mixture, the Mo^{5+} cation concentration has increased again, and it has returned to the state prior to annealing. Simultaneously, the binding energy of the Mo_{3d} level returned to that in the double perovskite state [28–32].

The temperature dependence of the electrical resistivity has been considered for the determination of the influence of dielectric interlayers between the $Sr_2FeMoO_{6-\delta}$ grains on the electric charge transfer mechanism in the SFMO-a,b,c samples. It has been found that the SFMO-a,b,c samples reveal three different types of the $\rho = f(T)$ behavior depending on the annealing mode. The SFMO-a sample exhibits metallic conduction; as the dielectric interlayers on the grain boundaries are absent, this charge transfer mechanism will not be considered in the following. The SFMO-b sample has a low electrical resistivity, $\rho = 0.063$ Ohm·cm at $T = 300$ K, decreasing with the temperature decreasing down to $T = 143$ K, with a subsequent rise in the temperature range from 143 down to 4.2 K. The SFMO-b sample has a higher resistivity, $\rho = 335$ Ohm·cm at $T = 10$ K and a clearly expressed semiconducting character of the conductivity with a negative temperature coefficient of the resistivity ($d\rho/dT < 0$), which indicates the presence of continuous dielectric interlayers (Fig. 2.6).

The linear dependence of the normalized resistivity ρ/ρ_0 on the temperature, plotted on the $\ln(\rho/\rho_0)$ vs. $\ln(T_0/T)$, $\ln(\rho/\rho_0)$ vs. $f(T^{-1/2})$ and $\ln(\rho/\rho_0)$ vs. $f(T^{-1/4})$ scales (ρ/ρ_0 is the normalized electrical resistivity, and ρ_0 is the electrical resistivity at $T_0 = 300$ K), indicates the presence of a tunneling barrier. In the temperature range from 300 to 263 K, the polythermal dependence of the normalized electrical resistivity is a linear function of the inverse temperature, and it is sufficiently well described by the power law $\ln(\rho/\rho_o) \propto T^{-1/4}$, indicating the hopping mechanism of the conductivity through a energy barrier (Fig. 2.7a).

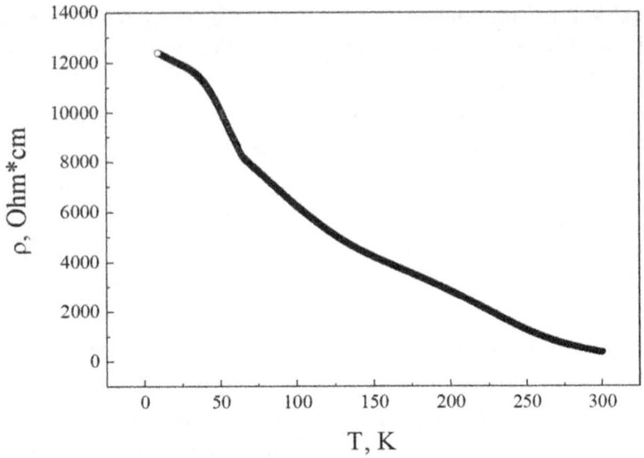

Fig. 2.6 Temperature dependences of the electrical resistivity of the SFMO-b sample with granular structure and dielectric shells

The presence of localized states in the dielectric shell influences the conductivity mechanism in the temperature range $T = 130–200$ K, where the law $\ln(\rho/\rho_0) \propto T^{-1/2}$ is observed (Fig. 2.7b). This power-law dependence denotes charge transfer in the sample by means of the electron tunneling between $Sr_2FeMoO_{6-\delta}$ metallic grains from one energy state to another through the $SrMoO_4$ energy barriers. In the framework of this model, the tunneling conductivity is determined by the energy barrier parameters, and it can be expressed as:

$$\rho \propto \exp\left((2/\hbar)[2m\phi]^{1/2}s\right), \tag{2.1}$$

where $\hbar$ is Planck's constant; m is the effective mass of the electrons; ϕ is the effective barrier height, s is the barrier width which is equal to the minimum length of electron hopping. In this case, the electroneutrality is violated upon electron hopping, and during the tunneling process the electron additionally needs to overcome an energy barrier formed by the Coulomb interaction. As a result, only electrons being thermally activated above the Coulomb energy barrier can move.

At lower temperatures (4.2–50 K), the best linearization satisfies the power law $\ln(\rho/\rho_0) \propto \ln(T_0/T)$ (Fig. 2.7c). When the phonon energy stops to influence the charge transfer, in the temperature range 4.2–50 K the electrons can suffer inelastic tunneling via separate conducting channels from one metallic grain to another through localized states.

As can be seen in Fig. 2.8, the experimental $\rho = f(T)$ dependences in an external magnetic field have a clearly expressed semiconducting character with a negative temperature coefficient of the resistivity ($d\rho/dT < 0$) in the entire magnetic field range $B = 0–10$ T, i.e. the total conductivity of the system lowers as the temperature decreases. The expression $d\rho/dT < 0$ is characteristic of the reduced double perovskites with a superstructural ordering of the cations in the presence of a barrier on the dielectric side.

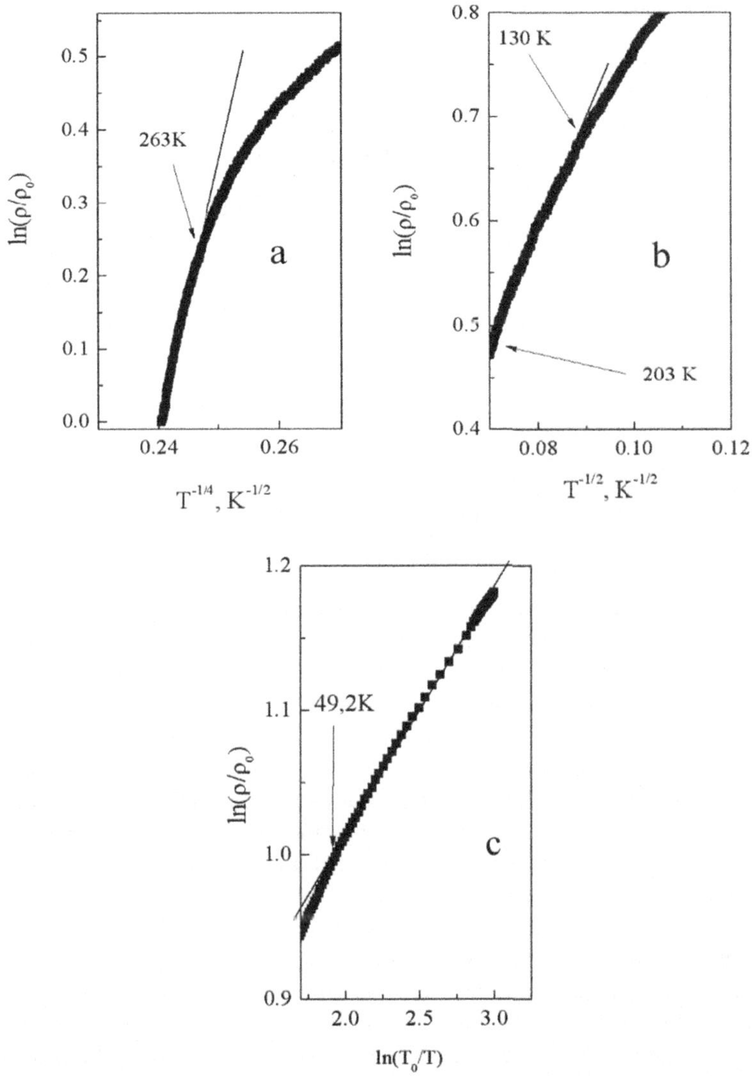

Fig. 2.7 Temperature dependences of the normalized electrical resistivity of the SFMO-b sample: (**a**) in the $\ln(\rho/\rho_0)$ vs. $T^{-1/4}$ coordinates; (**b**) in the $\ln(\rho/\rho_0)$ vs. $T^{-1/2}$ coordinates; (**c**) in the $\ln(\rho/\rho_0)$ vs. $\ln(T_0/T)$ coordinates

As it has been indicated above, the polythermal dependence of the normalized electrical resistivity was determined in the high-temperature region, in the temperature range from 250 to 300 K. This dependence is a linear function of the inverse temperature, and it is well described by the power law $\ln(\rho/\rho_0) \propto T^{-1/4}$, typical of the hopping conductivity through tunneling barriers (Fig. 2.9a). The application of magnetic fields $B = 5$ T and $B = 10$ T lowers the electrical resistivity of the sample

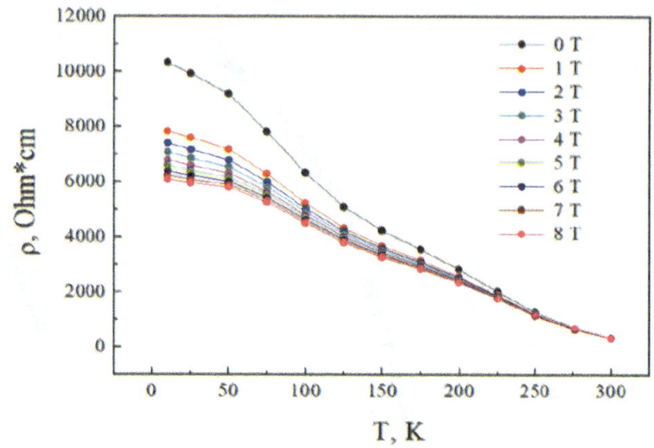

Fig. 2.8 Temperature dependences of the electrical resistivity of the granular SFMO-b sample at different values of applied magnetic field

in the entire temperature range studied, and it does not change the semiconducting type of the conductivity. The application of a magnetic field only shifts the onset of the charge transfer hopping mechanism toward lower temperatures, with the temperature coefficient of the resistivity keeping its sign (Fig. 2.9b,c).

It should be noted that the normalized electrical resistivity rises by an order of magnitude with decreasing temperature, $\rho(4.2)/\rho(300) \approx 20$, which indicates a clearly expressed semiconductor conductivity. The electrical resistivity of the SFMO-b sample in the low-temperature region starts with a sharp increase, which is caused by the appearance of additional scattering centers of the free charge carriers. The centers appear both due to the weak localization effects and the electron-electron interaction. Also, the phase boundaries possess an ability to increase the scattering in the low-temperature region. The phase boundaries influence the spin-dependent transport and the tunneling current in the conductivity between the $Sr_2FeMoO_{6-\delta}$ magnetic granules.

The linearized experimental dependences $\rho = f(T)$, constructed in the low-temperature region at different magnetic fields, indicate an increase of the tunneling current fraction with the further temperature decrease. It has been found that in the magnetic field the temperature range, where the $\ln(\rho/\rho_o)$ vs. $\ln(T/T_0)$ dependence is linear, is expanded (Fig. 2.10). This fact evidences an increase of the contribution of the inelastic resonance tunneling of electrons through the individual conducting channels from one metallic grain to another on localized states in the dielectric shell [29, 31, 32].

2.5 Current-Voltage Characteristics

Investigations of the current-voltage characteristics (CVC) play an important role in the determination of the conduction mechanisms in granular SFMO structures.

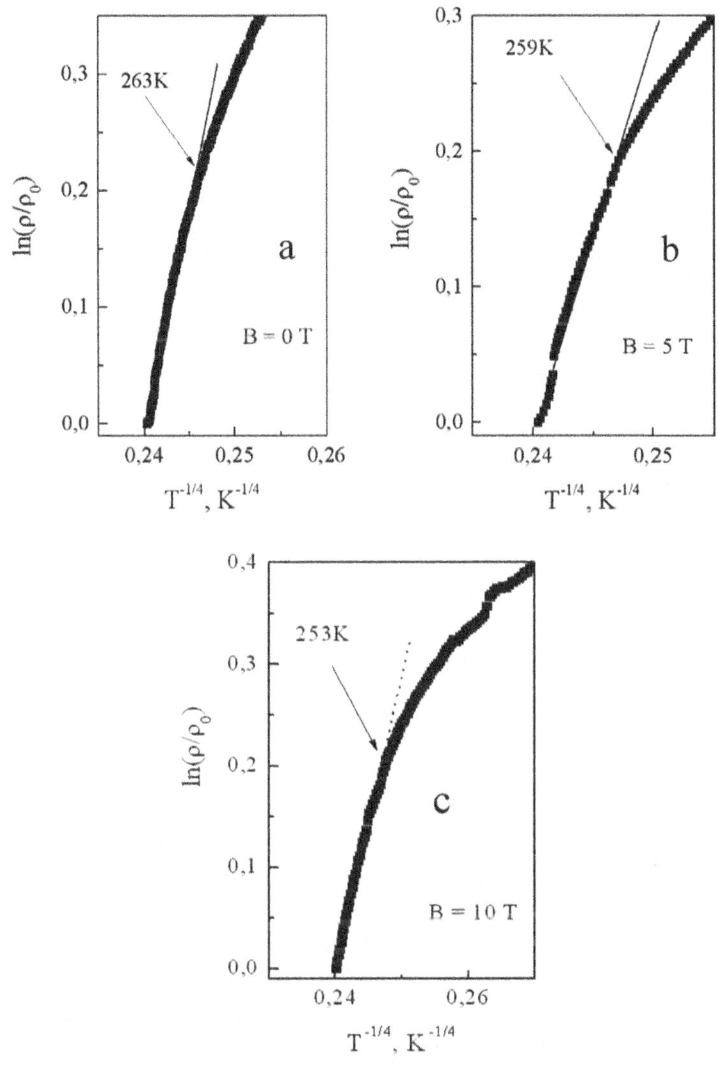

Fig. 2.9 Temperature dependences of the normalized electrical resistivity of the SFMO-b sample in the $\ln(\rho/\rho_0)$ vs. $T^{-1/4}$ coordinates measured at different values of applied magnetic field

The results of the investigations of the electrical properties of the nanocomposites described above were obtained at constant currents lower than 10^{-5} A. These currents were sufficient for the reliable measurements of the voltage drop between potential electrodes. The studies of CVCs of the composites were necessary for the determination of the $SrMoO_4$ dielectric shells thickness, as shown in Fig. 2.11.

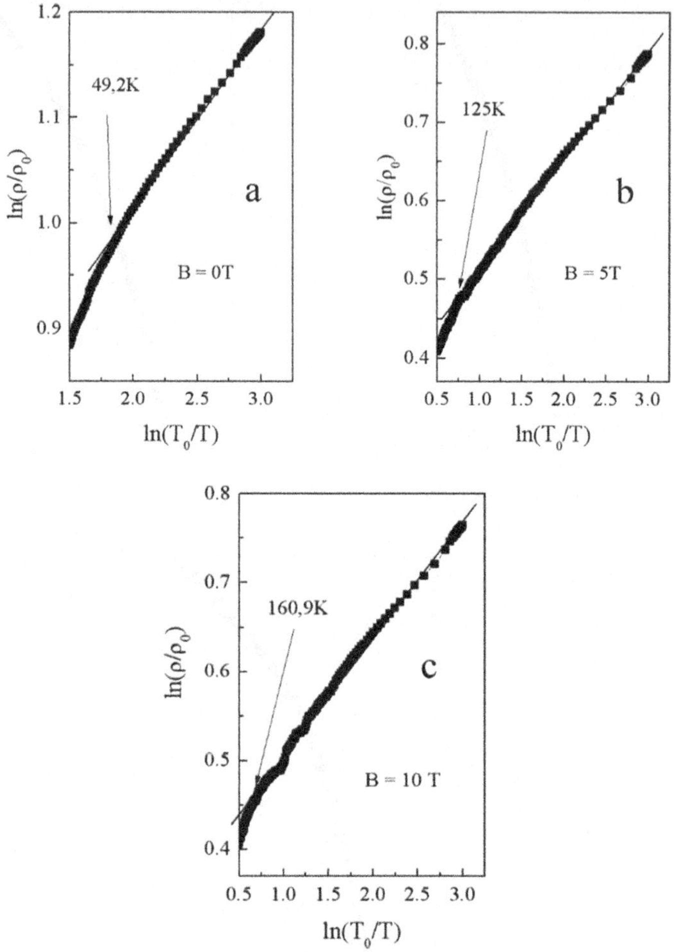

Fig. 2.10 Temperature dependences of the normalized electrical resistivity of the SFMO-b sample in the ln(ρ/ρ_0) vs. ln(T_0/T) coordinates measured in an external magnetic field of 10 T

First, one should point out that CVCs are symmetric at the negative and positive voltages. Magnetic fields up to 10 T do not influence the *I-U* dependences. The latter have a linear character up to the voltage values where a sharp leap occurs. In the temperature range from 10 to 315 K, the leap voltage changes insignificantly, from 3.6 to 4.0 V at room temperature. It is quite evident that such a sharp current leap is related with the electric breakdown in the $SrMoO_4$ dielectric shells between the conductive $Sr_2FeMoO_{6-\delta}$ grains. Based on the well-known data on complex metallic oxides (to which belongs $SrMoO_4$), one could assert that their breakdown

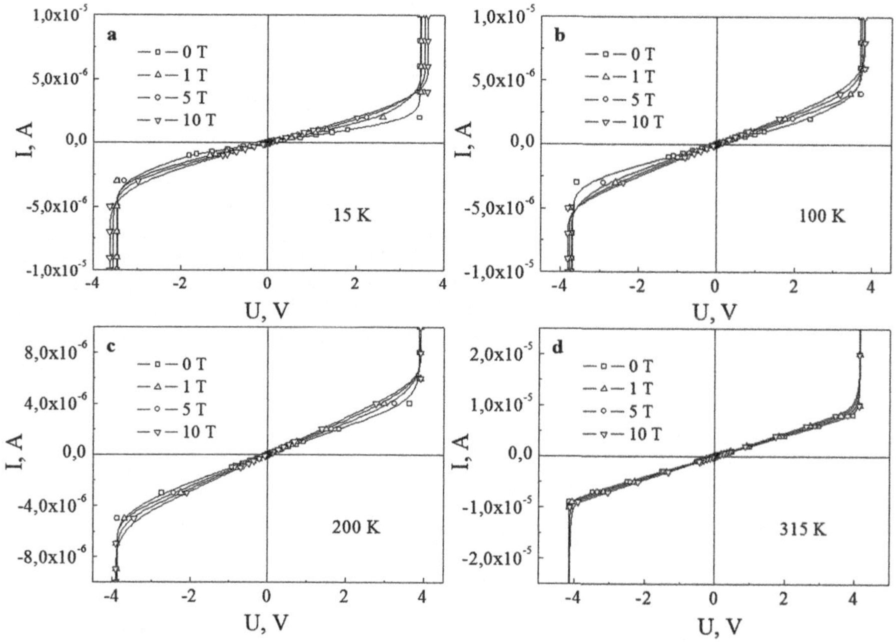

Fig. 2.11 Current-voltage characteristics of the SFMO-b granular structure, measured in magnetic fields at T = (**a**) 15 K; (**b**) 100 K; (**c**) 200 K; (**d**) 315 K

voltage U_{br} is about 20 kV/mm. Using the minimal U_{br} value for the breakdown voltage of 4 V, we get a dielectric interlayer thickness between the SFMO grains of about 20 nm.

This means that the thickness of the shell around every grain will be about 10 nm, whereas the average SFMO grain size will decrease down to 50 nm. A characteristic feature of this dependence is the reversibility: with decreasing voltage the CVCs are completely recovered. This effect is similar to the electrical breakdown in semiconductor p–n junctions. However, its nature is different: we consider that the breakdown mechanism is determined by the impact ionization in the $SrMoO_4$ layers in a strong electric field. The knocked-out electrons contribute to the electrical conduction, which makes the interlayer between the SFMO grains transparent.

After the realization of the above mentioned investigations, we carried out direct observations of the dielectric shells around SFMO granules by means of the field emission SEM technique (Fig. 2.12). This is a direct verification of the CVC investigations carried out earlier, whose result was that the average thickness of the $SrMoO_4$ dielectric shells is on the order of 10 nm. This in turn promotes the electron tunneling between the metallic SFMO granules.

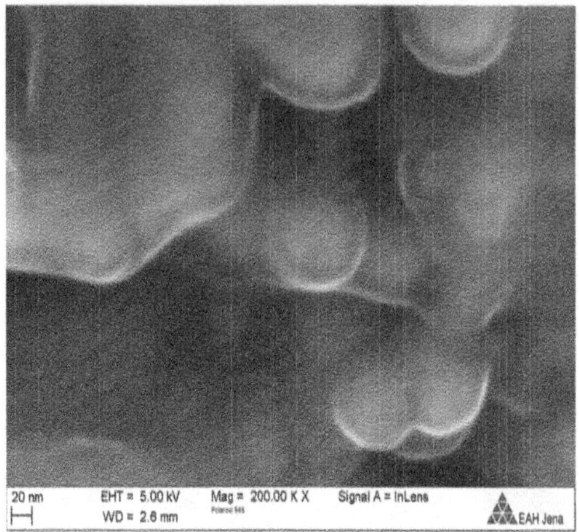

Fig. 2.12 Field emission SEM image of the SFMO sample with dielectric shells

2.6 Magnetoresistance Effects

The interest to investigations of the magnetoresistance in granular composites on the base of nanodispersed particles of magnetic materials embedded in a dielectric matrix has arisen in connection with the negative MR effect observed there, achieving several tens of percent in magnetic fields above 1 T. This effect is concerned with the spin-dependent tunneling of conduction electrons between nanoparticles, whose magnetic moments are aligned along the magnetic field, and it is observed in sufficiently strong magnetic fields [33, 34].

First measurements of the MR were carried out at room temperature in low transverse magnetic fields below 0.15 T (Fig. 2.13). The reason for a negative MR is obviously the appearance of the tunneling MR. Since the electric charge transfer is realized between the metallic SFMO grains which are separated by dielectric interlayers, the nonlinear character and positive sign of the MR in low magnetic fields is caused by the dominating activation character of the charge transfer. The value of the positive MR achieved 30% in the fields of about 0.05 T.

Dependences of a similar shape were obtained in Ref. [35] in the investigation of islanded composite films of the ferromagnetic-metal type. The respective positive MR in low magnetic fields achieved just 1–2%. The result obtained in the present work evidences the extreme sensitivity of the $Sr_2FeMoO_{6-\delta}$ - $SrMoO_4$ composite to the influence of the super-weak magnetic fields, which makes it possible to detect fields down to 10^{-9} T.

MR investigations in strong magnetic fields are presented in Fig. 2.14. The $\rho(T)$ curves in the entire temperature and magnetic field range have a clearly expressed semiconducting character (Fig. 2.8). This is natural for the metal-dielectric composites, when a dielectric plays the role of a potential barrier between metallic grains. The conductivity increases with the magnetic field, since the probability of the spin-polarized free electron tunneling increases. It has been revealed that a negative MR

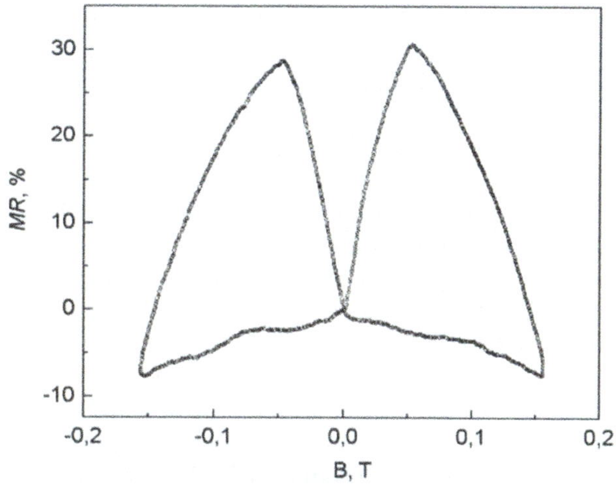

Fig. 2.13 Magnetic field dependence of the magnetoresistance for the SFMO sample with dielectric shells taken at $T = 300$ K

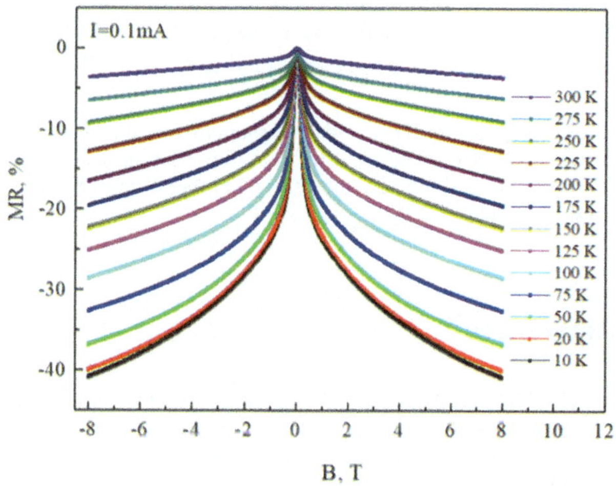

Fig. 2.14 Magnetic field dependence of the MR at various temperatures for the SFMO-b sample

is observed with increasing magnetic field induction. Therefore, it has been shown that a negative MR dominates in a broad range of temperatures and magnetic fields, and the maximal MR value equals 43% in a field of 8 T at 10 K.

2.7 Conclusions

In the present work, the $Sr_2FeMoO_{6-\delta}$ ferrimagnetc nanopowder was synthesized for the first time using a modified citrate-gel technique. The samples, revealing the metallic conductivity type according to the temperature dependence of the resistivity, have been obtained by the compaction of the powder. After a heat treatment in an argon atmosphere, the SFMO particles were covered with $SrMoO_4$ dielectric shells. The conductivity type in the obtained nanocomposite changes from metallic to semiconducting in the temperature range from 4.2 to 300 K and in magnetic fields up to 10 T. Investigations of the field dependences of the MR have shown that the negative MR dominates in the nanocomposite. The appearance of this MR is concerned with the spin-polarized charge carrier tunneling through the dielectric shells. The maximal MR value achieved 43% in a field of 8 T at 10 K. The studies of current-voltage characteristics of the composites were carried out for the determination of the $SrMoO_4$ dielectric shell thickness. These characteristics are symmetrical at negative and positive voltages; magnetic fields up to 10 T do not influence the *I-U* dependences; the latter have a linear character up to the voltage values at which a sharp leap of the current takes place. It is related with the electrical breakdown of the dielectric shells which has a reversible nature. The breakdown mechanism is governed by the impact ionization of the dielectric shells in strong electric fields, due to which the electrons are knocked out from $SrMoO_4$. These electrons contribute to the electrical conduction, which makes the border between the SFMO grains transparent. From the measured breakdown voltage, the dielectric shells thickness has been concluded to be about 10 nm.

Acknowledgements This work was supported by the Deutsche Forschungsgemeinschaft (DFG) Project No. MA 2359/30-1, by the European project H2020-MSCA-RISE-2017-778308 – SPINMULTIFILM, and by the FCT of Portugal through the project I3N/FSCOSD (Ref. FCT UID/CTM/50025/2013).

References

1. Serrate D, De Teresa JM, Ibarra MR (2007) Double perovskites with ferromagnetism above room temperature. J Phys Condens Matter 19:1
2. Topwal D, Sarma DD, Kato H, Tokura Y, Avignon M (2006) Structural and magnetic properties of $Sr_2Fe_{1+x}Mo_{1-x}O_6$ ($-1<x<0.25$). Phys Rev B 73:0944191
3. Chan TS, Liu RS, Hu SF, Lin JG (2005) Structure and physical properties of double perovskite compounds Sr_2FeMO_6 (M = Mo, W). Mater Chem Phys 93:314
4. Rager J, Zipperle M, Sharma A, MacManus-Driscoll JL (2004) Oxygen Stoichiometry in Sr_2FeMoO_6, the determination of Fe and Mo valence states, and the chemical phase diagram of $SrO - Fe_3O_4 - MoO_3$. J Am Ceram Soc 87:1330
5. Kanchana V, Vaitheeswaran G, Alouani M, Delin A (2007) Electronic structure and x-ray magnetic circular dichroism of Sr_2FeMoO_6: Ab initio calculations. Phys Rev B 75:220404
6. Szotek Z, Temmerman WM, Svane A, Petit L, Winter H (2003) Electronic structure of half-metallic double perovskites. Phys Rev B 68:104411
7. Sarma DD (2001) A new class of magnetic materials: Sr_2FeMoO_6 and related compounds. Curr Opin Solid State Mater Sci 5:261
8. Kobayashi KI, Kimura T, Sawada H, Terakura K, Tokura Y (1998) Room-temperature magnetoresistance in an oxide material with an ordered double-perovskite structure. Nature 395:677
9. Jurca B, Berthon J, Dragoe N, Berthet P (2009) Influence of successive sintering treatments on high ordered Sr_2FeMoO_6 double perovskite properties. J Alloys Compd 474:416

10. Fang T-T, Lin J-C (2005) Formation kinetics of Sr_2FeMoO_6 double perovskite. J Mater Sci 40:683
11. Klencsár Z, Németh Z, Vértes A, Kotsis I, Nagy M, Cziráki Á, Ulhaq-Bouillet C, Pierron-Bohnes V, Vad K, Mészáros S, Hakl J (2004) The effect of cation disorder on the structure of Sr_2FeMoO_6 double perovskite. J Magn Magn Mater 281:115
12. Ibarra MR, de Teresa JM (1998) Colossal magnetoresistance in manganese oxide perovskites. J Magn Magn Mater 177:846
13. Fix T, Barla A, Ulhaq-Bouillet C, Colis S, Kappler JP, Dinia A (2007) Absence of tunnel magnetoresistance in Sr_2FeMoO_6-based magnetic tunnel junctions. Chem Phys Lett 434:276
14. Yuan CL, Zhu Y, Ong PP, Ong CK, Yu T, Shen ZX (2003) Grain boundary effects on the magneto-transport properties of Sr_2FeMoO_6 induced by variation of the ambient H_2–Ar mixture ratio during annealing. Physica B Condens Matt 334:408
15. Raittila J, Salminen T, Suominen T, Schlesier K, Paturi P (2006) Nanocrystalline Sr_2FeMoO_6 prepared by citrate-gel method. J Phys Chem Solids 67:1712
16. Li XH, Sun YP, Li WJ, Ang R, Zhang SB, Zhu XB, Song WH, Dai JM (2008) Size dependence of electronic and magnetic properties of double perovskite Sr_2FeMoO_6. Solid State Commun 145:98
17. Yuan CL, Wang SG, Song WH, Yu T, Dai JM, Ye SL, Sun YP (1999) Enhanced intergrain tunneling magnetoresistance in double perovskite Sr_2FeMoO_6 polycrystals with nanometer-scale particles. Appl Phys Lett 75:3853
18. Nakayama S, Nakagawa T, Nomura S (1968) Neutron diffraction study of $Sr_2(FeMo)O_6$. J Phys Soc Jpn 24:219
19. Kumar D, Kaur D (2011) Exchange biasing in SFMO/SFWO double perovskite multilayer thin films. J Alloys Compd 509:7886
20. Sui Y, Wang XJ, Quan ZN, Liu ZG, Miao JP, Cheng JG, Huang XQ, Lu Z, Su WH, Ong CK (2005) Influence of doping Al on the magnetoresistance of polycrystalline Sr_2FeMoO_6. J Magn Magn Mater 290–291:1089
21. Aloysiusa RP, Dhankhar M, Kotnala RK (2013) Enhanced low field magnetoresistance in $Sr_2Fe_{1-x}Ag_xMoO_6$ double perovskite system. J Alloys Compd 574:335
22. Wang J-F, Hu B, Zhang J, Gu Z-B, Zhang ST (2015) Enhanced low-field magnetoresistance in organic/inorganic glycerin. J Alloys Compd 621:131
23. Cibert J, Bobo J-F, Lüders U (2005) Development of new materials for spintronics. CR Physique 6:977
24. Huang YH, Lindén J, Yamauchi H, Karppinen M (2005) Large low-field magnetoresistance effect in Sr_2FeMoO_6 homocomposites. Appl Phys Lett 87:072510
25. Tai LW, Lessing PA (1992) Modified resin-intermediate processing of perovskite powders. Part II. Processing for fine, nonagglomerated Sr-doped lanthanum chromite powders. J Mater Res 7:511
26. Yarmolich M, Kalanda N, Demyanov S, Terryn H, Ustarroz J, Silibin M, Gorokh G (2016a) Influence of synthesis conditions on microstructure and phase transformations of annealed $Sr_2FeMoO_{6-\delta}$ nanopowders formed by citrate-gel method. Beilstein J Nanotechnol 7:1202
27. Yarmolich M, Kalanda N, Demyanov S, Fedotova J, Bayev V, Sobolev NA (2016) Charge ordering and magnetic properties in nanosized $Sr_2FeMoO_{6-\delta}$ powders. Phys Status Solidi B Basic Solid State Phys 253:2160
28. Fang T-T, Lin J-C (2005) Formation kinetics of Sr_2FeMoO_6 double perovskite. J Mater Sci 40:683
29. Yarmolich MV, Kalanda NA, Demyanov SE, Silibin MV, Gorokh GG (2016) Synthesis and study of strontium ferromolybdate nanopowders with high degree of superstructural ordering for spintronics. TKEA 2-3:41
30. Kovalev LV, Yarmolich MV (2014) The influence of synthesis conditions on structural and magneto-transport propertis of $Sr_2FeMoO_{6-\delta}$. Proc Natl Acad Sci Belarus (physical-technical sciences series) 1:46

31. Yarmolich MV, Kalanda NA (2016) Charge ordering in a compound of $Sr_2FeMoO_{6-\delta}$, synthesized by citrate-gel method. Supplement to the journal "Proc National Acad Sci Belarus" (physical-technical sciences series) 4:80
32. Yarmolich MV, Kalanda NA, Demyanov SE, Hurski LI, Kovalev LV, Galyas AI (2016) Magnetism in the nanosized strontium ferromolybdate. Doklady BGUIR 3:63
33. Gittlman JL, Goldstain Y, Bozovski S (1972) Magnetic properties of granular nickel films. Phys Rev B 5:3609
34. Milner A, Gerber A, Groisman B, Karpovsky M, Gladkikh A (1996) Spin-dependent electronic transport in granular ferromagnets. Phys Rev Lett 76:475
35. Boltaev AP, Pudonin FA, Sherstnev A (2011) Specific features of the magnetoresistance in multilayer systems of magnetic nanoislands in weak magnetic fields. Phys Solid State 53:950

Chapter 3
Peculiarities of Formation and Characterization of SiO_2/Si Ion-Track Template

Egor Kaniukov, Victoria Bundyukova, Maksim Kutuzau, and Dzmitry Yakimchuk

Abstract A simple way of forming porous templates of silicon dioxide using ion-track technology was demonstrated. Investigations of the dependence of the characteristics, both of a single pore and of the entire template on the etching time and concentration of etchant were carried out. The possibility of high-precision control of the parameters of porous templates through the post-processing of SEM images in the ImageJ software was demonstrated. An express method for estimating template parameters using ellipsometry was proposed.

Keywords Nanotechnology · Templates · Swift heavy ion track technology · ImageJ · Ellipsometry

3.1 Introduction

In the last years a large interest to investigations of various nanosized objects was developed in concern with their exhibition of interesting characteristic properties, being indistinctive for materials with macro- and microscopic properties. The simplest and logical method of creation of nanomaterials is the use of nanoporous matrices (templates), as natural conditions of self-organization of nanosized objects are realized in the pores. Using of the pores makes it possible to obtain nanoparticles and nanofibres of semiconductors [1], metals [2], multilayers [3] of metals, etc. Some structures of this kind possess a sensitivity to light, humidity, organic substances, and could be used as sensors [4]. Depending on the nature of the material

---

E. Kaniukov (✉)
Scientific-Practical Materials Research Centre, NAS of Belarus, Minsk, Belarus

Institute of Chemistry of New Materials, NAS of Belarus, Minsk, Belarus
e-mail: ka.egor@mail.ru

V. Bundyukova · M. Kutuzau · D. Yakimchuk
Scientific-Practical Materials Research Centre, NAS of Belarus, Minsk, Belarus

A. Maffucci, S. A. Maksimenko (eds.), *Fundamental and Applied Nano-Electromagnetics II*, NATO Science for Peace and Security Series B: Physics and Biophysics, https://doi.org/10.1007/978-94-024-1687-9_3

placed in pores, the obtained nanostructures could be used for a construction of resistors, capacitors, diodes and inductors [5]. At present, porous Al_2O_3, TiO_2, polycarbonate and polyimide membranes are used for a creation of nano-objects [6–13]. Still, from the point of view of an implementation of nanosystems on the base of silicon in modern technological processes, most prospective is the use of templates on the base of silicon oxide. Pores in a-SiO_2 layer can be obtained with the implementation of the swift heavy ion (SHI) track technology, which includes irradiation with the SHIs and chemical etching [9].

The a-SiO_2 porous matrices can be used as a form-building basis for realization of nanostructures. It makes possible to obtain arrays of nanostructures with different morphologies: particles, rods, tubes, dendritic structures [3, 14–16], which can be useful for creation of new types of electronic devices, sensors, systems of biological control and etc. [10, 17]. With such approach, it is important to control such templates parameters as height, external and internal pore diameters and their density. For this purpose, various software solutions can be used. In particular, one of the most effective methods is using of ImageJ software in analysis of scanning electron microscopy images. This method allows to determine a wide range of pore parameters for obtaining a large amount of statistical information. Therefore, in this paper we demonstrate the possibility of surface analysis by SEM with ImageJ post-processing for SiO_2/Si ion-track template parameters control.

From the previous works [18–24] it is well known that the process of chemical etching in solutions of hydrofluoric acid of swift heavy ion tracks in a-SiO_2 depends on the energy of electronic loss $(dE/dx)_e$ and has a threshold character. At the $(dE/dx)_e$ value less than 1.5 $keV \cdot nm^{-1}$, pores formation is impossible. At the $(dE/dx)_e$ values in the range $\sim$ 1.5–4 $keV \cdot nm^{-1}$ the etching process is possible, but obtained pores are characterized by various diameters and angles of the etching cone, as well as by modest etching efficiency (not all the ion tracks are transformed into pores). The energy value of about 4 $keV \cdot nm^{-1}$ is a necessary condition for the obtaining of pores with a uniform distribution of forms and dimensions with the etching efficiency $\sim$ 100%. At the electronic loss energies of more than 4 $keV \cdot nm^{-1}$, the pore diameters depend on the $(dE/dx)_e$ value and their sizes increase with the rise of this loss at the same etching conditions.

A systematic investigation of results of the ion track etching in the a-SiO_2 layer, obtained at the irradiation by ^{132}Xe, ^{152}Sm, ^{197}Au and ^{238}U ions with $(dE/dx)_e$ values in the range from 16 to 23 $keV \cdot nm^{-1}$ made it possible to introduce the new threshold value $(dE/dx)_e > 15$ $keV \cdot nm^{-1}$. With that the pore diameters have constant size independently on the value of electronic loss energy [25]. Results of the present investigation make it possible to forecast obtaining of the a-SiO_2-based porous templates with a pre-determined geometrical parameters, independently of the irradiating ion with a high precision when silicon oxide layer thikness is known.

On the base of these we demonstrate the possibility of restoring information about the main pore parameters in the silicon oxide layer of ion track SiO_2/Si templates using only ellipsometry, without additional control techniques. The obtained results are compared with the SEM-microscopy data, and the limitations imposed on the proposed method are discussed.

3.2 Experimental Procedures

In the present work, (100) n-type silicon wafers (boron-doped with resistivity of 12 $\Omega \times$ cm) prepared by the Czochralski method and chemically polished on one side were used. Amorphous a-SiO_2 layers were thermally grown by pyrogenic oxidation at 1050 °C. The a-SiO_2 film thickness of 714 nm was controlled by laser interferometry.

Irradiation of the samples was carried out at the irradiation facilities of the GSI Helmholtz Centre for Heavy Ion Research (Darmstadt, Germany). The samples were bombarded at normal incidence with ^{152}Au, ions with an energy of 350 MeV. The ion energy provided the guaranteed pulling through the silicon dioxide layer with a slight change of electronic loss energy $(dE/dx)_e$. The irraditation fluences were chosen to minimize the spatial overlap of the individual conical pores appearing after etching of latent tracks in a-SiO_2. Values of the fluences were in the range from 10^7 to 5×10^8 cm^{-2}.

The heavy ion irradiated samples were cleaned in solution of 56% HNO_3 at 80 °C during 30 s, rinsed in deionized water and dried in nitrogen flow. Subsequently, the samples were etched in 0.7, 1.4, 2.8 and 5.6 vol.% aqueous hydrofluoric acid (HF) solutions for different times at room temperature. After etching, the samples were immediately washed in deionized water and dried in nitrogen flow. All samples irradiated with a given ion species were etched concurrently.

Morphology of the porous samples was characterized by field emission scanning electron microscopy (SEM), using a JEOL JSM-7000F microscope. Image analysis was performed with ImageJ software with ImageJ2 version. Control of the a-SiO_2 layers thicknesses was performed by ellipsometry. Measurements were carried out using a spectroscopic ellipsometer with configuration: instrument type – rotating analyzer with compensator; working spectral range 240–950 nm; angle of incidence 60°; achromatic microspot of 200 microns. Before the measurement, the instrument was calibrated on reference samples – substrate of SiO_2/Si without pores. The spectra of the angles of the polarization ellipse of the reflected radiation $\Psi(\lambda)$, $\Delta(\lambda)$ were measured. The thickness of the porous layer was found from the spectra by solving the minimization problem [26].

3.3 Silicon Oxide Templates Formation by the SHI Tracks Technology

Hydrofluoric acid is normally used due to its wide implementation in silicon-based technological processes for the production of integrated circuits on silicon substrates [27]. In a general case, the chemical etching of a-SiO_2 in aqueous solutions of HF takes place according to the following reaction:

$$SiO_2 + 6HF \rightarrow H_2SiF_6 + 2H_2O \tag{3.1}$$

The following factors influence the etching rate of irradiated and non-irradiated SiO_2: HF concentration; temperature; mechanical stirring during the etching process [27, 28], and finally the density and formation process of the etched SiO_2 [29].

The selective etching process of highly-defective areas in the SHI irradiated SiO_2 layer is related to the difference of etching rate of the damaged and undamaged material. V_B is the rate of dissolution of the undamaged material, whereas V_T is the etching rate along the SHI track. The formation, by etching, of a pores at the latent SHI track is accompanied by a decrease of the oxide layer thickness *(L)*. This is due to the simultaneous etching of damaged and undamaged material. The etching process leads to an increase of the upper diameter of the cone (*D*) and the lower diameter (*d*) together with a decrease of *L* (Fig. 3.1a).

The knowledge of V_T and V_B rates makes it possible to determine the main pore parameters as diameter on the surface *D*, height *l* and half-angle of the etching cone, θ. These parameters can be described at any given moment of time by the following equations:

$$l = (V_T - V_B)\, t, \tag{3.2}$$

$$D = 2V_B t[(V-1)/(V+1)]^{1/2}, \tag{3.3}$$

$$\theta = arcsin\,(1/V), \tag{3.4}$$

$$D = 2\,(L_{SiO2} - L)\sqrt{\frac{1 - sin\theta}{1 + sin\theta}}, \tag{3.5}$$

$$d = D - 2L \cdot tg\theta, \tag{3.6}$$

where the ratio $V = V_T / V_B$ is usually a constant [30, 31], which determines the pore shape, obtained during the etching process. Usually this value is in the range of $2 < V < 4$ [32]. The value $V > 1$ is a condition of the transformation of the SHI track into pore.

The typical SEM images of the surface of SiO_2/Si ion track template and cross-sectional views of the individual pore in the silicon oxide layer at different etching time are shown in Fig. 3.1b, c, d, e, f, g, h. Treatment of pre-irradiated SiO_2/Si substrates in hydrofluoric acid at a concentration of 1% and at a temperature of 20 ± 2 °C for 35 min (Sample 1), 55 min (Sample 2) and 90 min (Sample 3) allowed to form pores in the SiO_2 layer. Analysis of SEM images make it possible to define the thickness of SiO_2 layer for Samples 1–3: $L_1 = 540$ nm, $L_2 = 470$ nm and $L_3 = 310$ nm, respectively. Proceeding from formulas (2–4), the values of the outer and bottom pore diameters were calculated as $D_1 = 264$ nm, $D_2 = 365$ nm, $D_3 = 603$ nm and $d_1 = 0$ (pore is not opened), $d_2 = 100$ nm, $d_3 = 400$ nm.

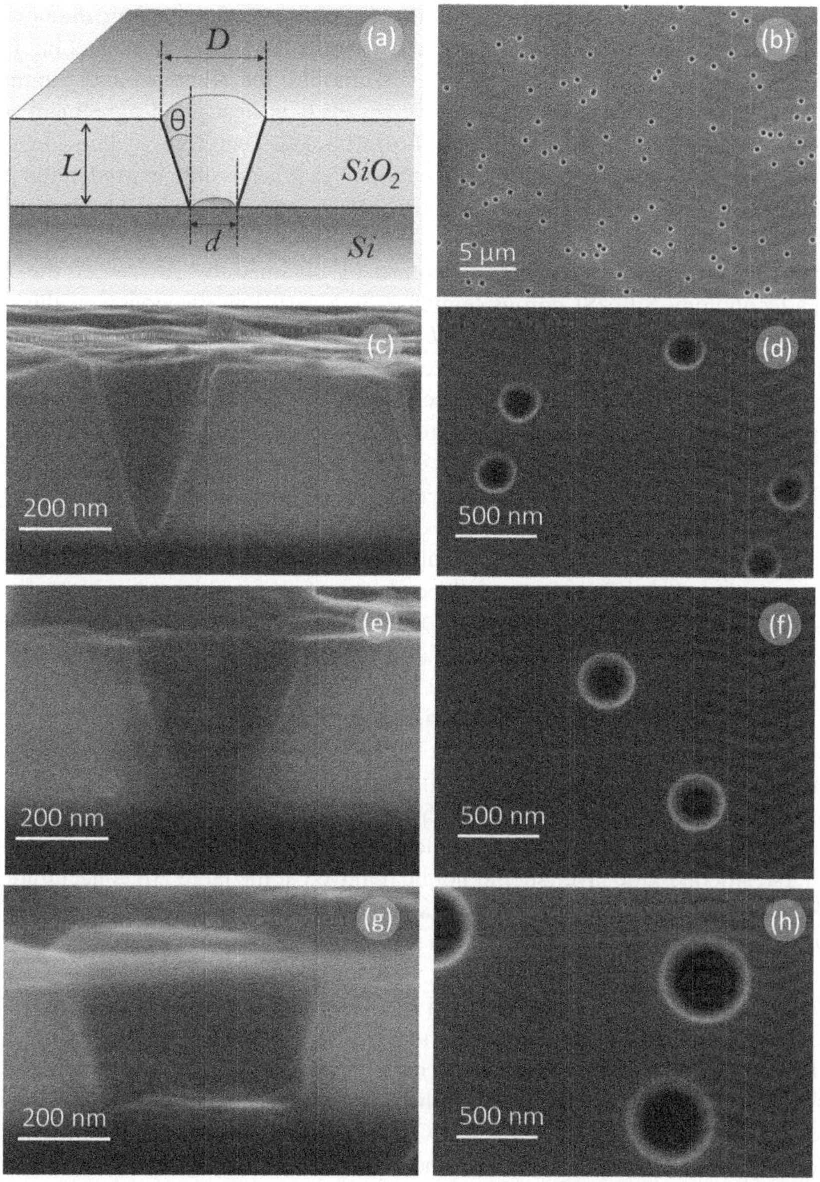

Fig. 3.1 Schematic representation of pore in SiO_2/Si ion track template (**a**): L – silicon oxide thickness after etching, D – upper and d – bottom pore diameters, θ – half-pore angle. SEM images of SiO_2/Si template surface (**b**, **d**, **f**, **h**) and cross-section (**c**, **e**, **g**) for Sample 1 (**c**, **d**), Sample 2 (**e**, **f**), Sample 3 (**g**, **h**)

It should be noted that the time evolution of all the pore characteristic dimensions has a linear character. This enables one to determine the values of the etching rates $V_T = 15.5 \pm 0.5$ nm/min, $V_B = 4.5 \pm 0.3$ nm/min, the adimensional parameter $V = 3.4 \pm 0.5$ and the angle $\theta = 17.0° \pm 2°$. It is in a good agreement with previous data for comparable quantities of $(\mathrm{d}E/\mathrm{d}x)_e$ [9]. Therefore, by knowing only the rate of the undamaged material dissolution (V_B) it is possible to predict the pore characteristic dimensions at any etching time. V_B is easily determined by measuring of SiO_2 thickness.

Another important characteristic feature of obtained by means of the SHI tracks technology SiO_2/Si porous templates is a very low dispersion of the pore characteristic dimensions (*D, d, l* and *L*) at any etching stage. An analysis of the standard deviation (σ) values of the pore characteristic dimensions shows that they do not exceed 6%. Such a small dispersion in the morphology of the obtained templates is an important asset for future integration of such templates in different application fields.

Study of the influence of HF concentration is interesting because of the fact that HF is a weak acid, which does not dissociate completely in aqueous solutions [33]. Therefore dimerization reaction can take place in HF solution and the dimers can lose one proton [28]. Since a dependence of the ionic composition on HF concentration in the solution could essentially influence the etching rates V_T and V_B, one needs to check, how this would affect the parameters of obtained pores.

A study of the dependence of pore parameters of the HF concentration was carried out by means of the SEM-investigations of the samples irradiated by ^{152}Sm ions with fluence $\Phi = 5 \times 10^8$ cm^{-2} at the HF solution concentration 0.7, 1.4, 2.8 and 5.6%. For all the samples, etching time was 18 min. The choice of the etching time is caused from the one hand by the fact that at low HF concentration pores should have dimensions which are sufficient for the investigation. From the other hand, the silicon oxide layer should not be removed at the large HF concentrations. Similar to the previous investigation, study of surface and cross-sections of a-SiO_2/Si systems with etched SHI tracks was carried out by means of SEM. Results of this study are presented in Fig. 3.2a.

Reaction of the latent tracks etching is a heterophase one and has a first order on concentration in the range from 0.7 to 5.6%. That is, the etching rate increases by factor of 2 at the twofold increase of HF concentration. Correspondingly, an increase of the concentration of the etching agent promotes a faster acquisition of porous templates on the base of the *a*-SiO_2 with required parameters. Still, with that the quality of obtained templates could be worse, which will be expressed in a larger dispersion of pores by their diameter values.

A correct comparison of the quality of the porous templates, which are obtained by using of HF with various concentrations, could be carried out only on templates with a similar parameters of pores. With the account of the fact that the $v(t)$ dependences are linear, it is obvious that the twofold increase of HF concentration would lead to the twofold increase of the etching rate. This made it possible for us to prepare a series of porous templates with similar characteristics with the use of following etching parameters: at $C_{HF} = 0.7\%$, $t = 36$ min; at $C_{HF} = 1.4\%$, $t = 18$ min; at $C_{HF} = 2.8\%$, $t = 9$ min; at $C_{HF} = 5.6\%$, $t = 4.5$ min.

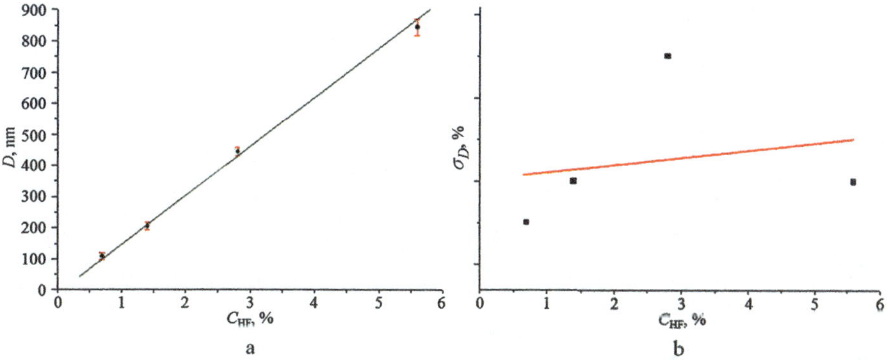

Fig. 3.2 Dependence of the pore outer diameter on HF solution concentration at the duration of etching 18 min (**a**). Dependence of the dispersion of upper pore diameter values (σ_D) on the HF concentration (straight line – linear fitting) (**b**)

In respect that the SEM analysis could be carried out more simple for the investigation of the samples surface, the mean square deviations σ_D of the upper pore diameters values were chosen as the parameter of comparison. The obtained values and their linear approximation are given in Fig. 3.2b. Data presented in the figure indicate the tendency of an increase of the dispersion of pore diameters values (σ_D) with the rise of HF concentration. Still, a small number of experimental points in the dependence does not allow to determine a type of the dependence and it requires an additional investigation. Nevertheless, the above mentioned tendency evidences the fact that the etching should be carried out in low-concentration HF solutions in order to obtain pores with a more homogeneous values of their diameters.

3.4 Post-processing of SiO_2/Si Ion-Track Template Images by ImageJ Software

The use of porous templates for controlled formation of nanostructures and nanosystems requires reliable method of pore size, shape, quantity per unit area (porosity), distance to nearest neighbors and porous matrix thickness monitoring. The scanning electron microscopy method allows to estimate these parameters, however, manual processing of obtained images is a very time-consuming task. The use of ImageJ software makes it possible to significantly simplify the process of obtaining detailed statistical information about templates surface. Taking that into account, the work describes in detail the methodology for analyzing of scanning electron microscopy images of SiO_2/Si ion-track templates surface to control pore size, shape, unit area and distance to the nearest neighbors. The method for estimating thickness of SiO_2/Si ion-track template based on processing of data about surface is also

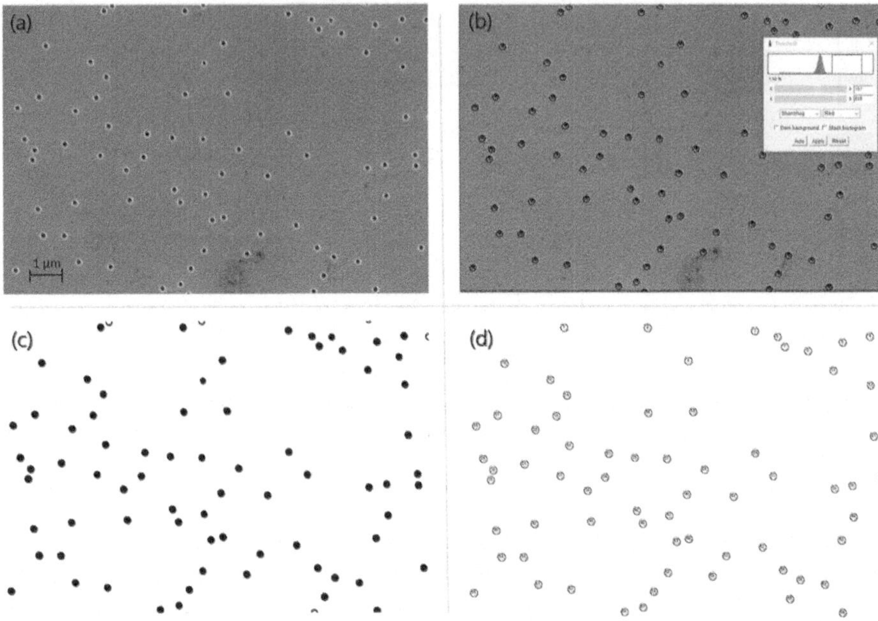

Fig. 3.3 (a) SEM image of porous SiO_2/Si template surface, (b) setting the threshold by using of ImageJ command *Threshold*, (c) image after binarization and (d) result of command *Analyze Particles* execution

proposed. It is important to note, that the techniques described in this paper not only allow to effectively control SiO_2/Si ion-track template pore parameters, but also suitable for other types of porous templates.

Surface analysis by ImageJ post-processing of SEM images was carried out on images of SiO_2/Si ion-track template surface, typical for the selected irradiation and etching modes (Fig. 3.3a).

From Fig. 3.3a it is seen that in connection with the random nature of the ion distribution in the irradiating beam, the pores in the oxidized silicon layer are stochastically distributed. Determination of pores characteristic parameters was carried out by using of *Analyze Particles (Analyze → Analyze Particles)* command of ImageJ software. Running the *Set Measurements* command *(Analyze → Set Measurements)* allows to set analyzed parameters. The selection of parameters in menu *(Measure → Set Measurements)*, *Analyze Particles* determines which statistical data about image (area and perimeter of selection, tilt angles, pore coordinates and etc.) will be displayed in a special window *Results*.

For efficient data processing, preliminary preparation of images by standard command of ImageJ package *Threshold (Image → Adjus → Threshold)* and image binarization *(Process → Binary → Make Binary)* are necessary. Using of *Threshold* command allows to interactively adjust upper and lower threshold values for segment of area of interest and image background (Fig. 3.3b). The threshold level is determined by analysis of histogram of current selection or full image if there is no selection. Setting of minimum and maximum exceptions ranges from different parameters after binarization (Fig. 3.3c) allows to extract information from binary image.

A more detailed image description can be obtained by using of BioVoxxel Toolbox. The *Extended Particle Analyzer* is based on ImageJ command *Analyze Particles* and allows to restrict particle analysis in accordance with many parametric descriptions of shape and angular orientations. The results output types are the same as for *Analyze → Analyze Particles*, namely: converted image and two-dimensional array of described above statistical data. It is possible to obtain information about the pores like area, extent perimeter, circularity, roundness, solidity, compactness, aspect ratio, Feret aspect ratio, ellipsoid angle (degree), max and min Feret, Feret angle and coefficient of variation by using of *Extended Particle Analyzer*. Information about function, which is realized by separate line, is shown in Fig. 3.4.

The output form of the results can be also set for working with them (display results, summarize, reset after analysis, clear results, add to manager) and for include/exclude of objects located on border of image and on area inside segmented regions (include holes, exclude edges).

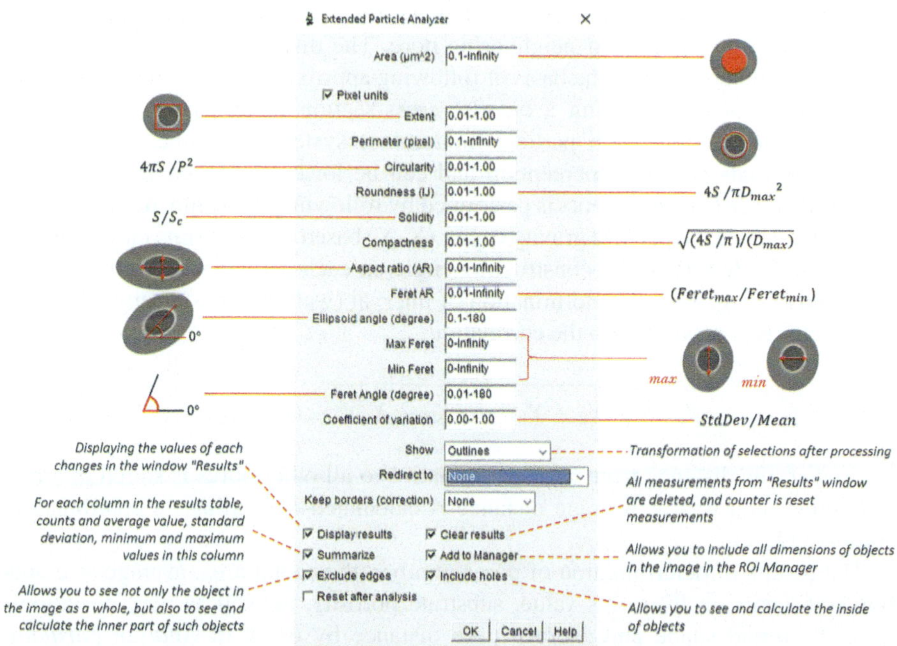

Fig. 3.4 Setting parameters in *Extended Particle Analyzer*, where S – area, S_c– area of a convex domain, P – perimeter, D_{max}– maximum diameter, *StdDev* – standard deviation of the area, *Mean* – mean value of the area

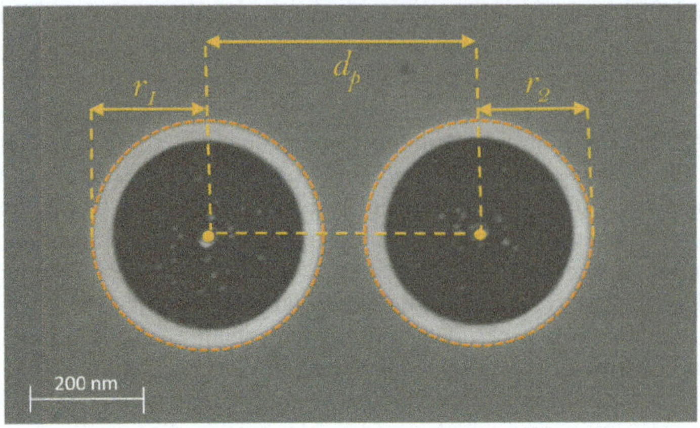

Fig. 3.5 Determination of pores radius and distance between pores by using of plugin *Nnd*

It is possible to obtain information about distance between the nearest pores with the plugin *Nnd (Plugins → Nnd),* which is added to built-in functions of ImageJ. As a result, the screen displays an image with pore contours and their numbering (Fig. 3.3d), as well as a window with table, which contains information about distance to the nearest neighboring pore. The distance between the nearest neighbors is determined on the basis of following approximations. In tightly-packed configuration of pores having a circular cross-section in 2D space, there are 6 neighbors surrounding each particle. In random system order the coordination number depends on visual perception and can be lower or higher. Estimation of particle distance to its neighbors is performed by following algorithm: determination of coordinate of each particle gravity center (X, Y) based on the results table of built-in plug-in *Analyze Particles* constructing circle on each particle with center (X, Y) and radius *r* (Fig. 3.5) → determination of interval (wall thickness) between a pair of particles (d_p) according to the correlation:

$$d_p = \sqrt{(Y_2 - Y_1)^2 + (X_2 - X_1)^2} - (r_1 + r_2) \tag{3.7}$$

Further, the distance from each pore relative to all other pores is stored in array and sorted, and finally, average distance is calculated depending on coordination number [34, 35].

The results of determination of pores number, their total area, average size and standard deviation from this value, substrate porosity, as well as pore deviations from the round shape and average pore distance by using of *Analyze Particles* command and *Nnd* plugin are given in Table 3.1.

On the basis of obtained results, the average value of upper pore diameter (*D*) – 0.39 μm is easily determined. The total area of image was 371.84 μm^2, it allows to calculate number of pores per unit surface – 1.85×10^7 cm^{-2}. The obtained value

Table 3.1 Basic pore parameters determined by using of ImageJ

Number of pores	Total pores area, μm^2	The average pore size, μm^2	Average standard deviation, %	Porosity, %	Pore deviation from round shape, %	Average distance between pores, μm
69	8,2	0,12	4	2,2	5	1,45

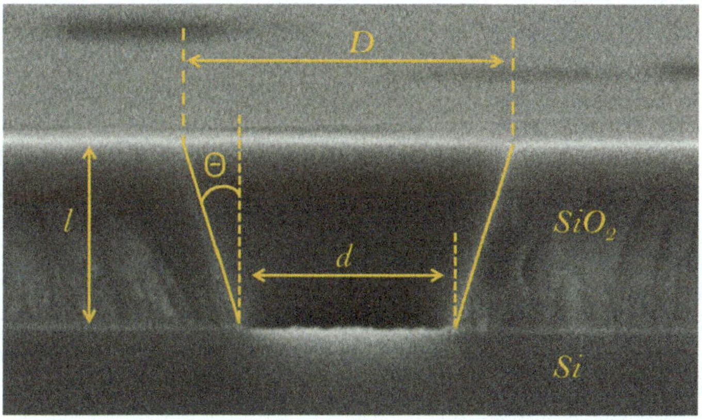

Fig. 3.6 SEM image of the SiO_2/Si ion-track template cross section with schematic representation of main pore parameters: D – upper pore diameter; l – silicon oxide thickness after etching; d – internal pore diameter; θ – half-pore angle

is lower than irradiation fluence (2×10^7 cm^{-2}) because selected parameters of command *Analyze Particles* were set so as to automatically exclude pores, which are partially outside the image. The presence of such pores is easy to see during comparing of SEM image and its processed version (Fig. 3.3a, d). The average distance to the nearest neighbor of each pore was 1.45 μm, it agrees well with theoretical calculations for the chosen fluence, given in [25]. It is important to note that standard deviation from average area value for each pore does not exceed 4%, and shape deviation from an ideally round is less than 5%. The small values of these parameters indicate the high quality of the SiO_2/Si templates obtained by ion tracks chemical etching.

Typically, the determination of pore parameters such as silicon oxide thickness after etching (l) and internal pore diameter (d) requires study of SiO_2/Si ion-track template cross section, which is time-consuming task. In addition, information on thickness of template obtained by using of SEM depends strongly on quality of prepared SiO_2/Si ion-track template cross section and orthogonally of sample cleavage plane electron beam. An example of SiO_2/Si ion-track template cross section with schematic representation of main pore parameters is presented in Fig. 3.6.

It should be noted that for l and d determination, the method for estimating of thickness of SiO_2/Si ion-track template based on information on upper pore diameters obtained during post-processing of SiO_2/Si ion-track template surface images can be used. It is known that etching parameters of irradiated SiO_2/Si substrates and characteristics of resulting pores are interrelated through a half-pore angle θ, which is determined by irradiation parameters. Earlier, in work [25] it was shown that θ at high energies of irradiating ions (electronic energy value $(dE/dx)_e > 16$ keV × nm^{-1}), pore characteristics do not depend on irradiating ion type, and the half-angles of etching cone are in the range of 17° ± 2°. Calculation of silicon oxide thickness after etching and internal pore diameter of SiO_2 can be performed by correlations (2–6).

During estimating of l and d, half angle was assumed equal to 17° and pore characteristic values determined as a result of calculations were $l = 0.257$ and $d = 0.231$ μm. It should be noted that information on θ is not known in advance, when irradiation energy of corresponding electronic loss energy value $16 \geq (dE/dx)_e \geq 4$ keV × nm^{-1}, but once obtained information on this parameter it can be extrapolated to all experiments of pore SiO_2/Si ion-track template etching with the selected energy and ion type by using of Eqs. (3.2, 3.3, 3.4, 3.5 and 3.6). Thus, the conducted studies visually demonstrate that using of ImageJ for processing of SEM images allows to obtain a large amount of statistical data on template parameters and significantly saves time during research.

3.5 Ellipsometry as an Express Method for Determining the Pore Parameters

Due to the development of ion-track technology, it became possible to produce porous templates with large areas, that are of interest for mass production of nanostructures. Given that the template parameters are often define the properties of the resulting nanostructures and nanosystems, a reliable method for non-destructive testing is needed for rapid control of template parameters. Such method could be ellipsometry, allowing for a single measurement to collect statistical information from a large area and to save time for certification. In order to adapt the ellipsometry method to control the parameters of ion-track patterns, the first experiments on the study of SiO_2/Si templates with low porosity were carried out. Using the standard model of the interaction of a polarized light beam with a layered structure of silicon oxide on silicon, the basic parameters of the pores were determined through mathematical transformations and subsequently were compared with the results of scanning electron microscopy.

Treatment of pre-irradiated SiO_2/Si substrates in hydrofluoric acid at a concentration of 1% and at a temperature of 20 ± 2 °C for 60 min allowed the formation of pores in the SiO_2 layer. The selectivity of the etching process was ensured by a higher etching rate of the irradiated (and therefore more highly defective) silicon oxide sites compared to the bulk silicon oxide etching rate. The surface of the SiO_2/Si ion-track template with individual pores in the silicon oxide layer are shown in Fig. 3.7a.

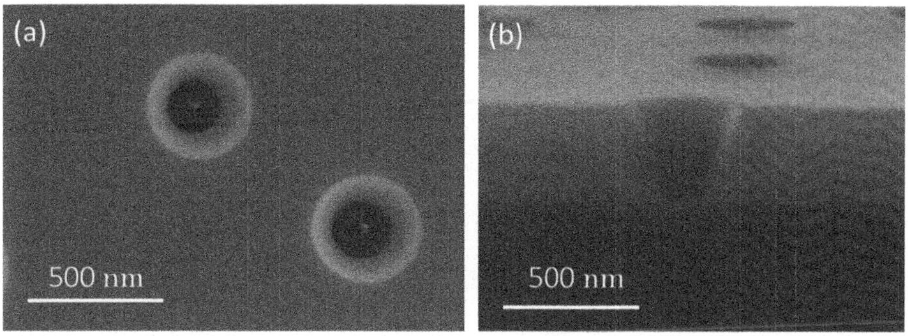

Fig. 3.7 SEM image of the SiO_2/Si ion-track template: individual pores (**a**); cross-section (**b**)

Table 3.2 Basic pore parameters determined

Number of pores	Total pores area, μm^2	The average pore size, μm^2	Average standard deviation, %	Porosity, %	Pore deviation from round shape, %	Average distance between pores, μm
78	12,2	0,16	4	4,1	7	0,98

Processing of SEM images using ImageJ made it possible to estimate the character of pore distribution over the surface, and also to establish the pore diameters on the surface – $D = 450 \pm 20$ nm. The results of determination of pores number, their total area, average size and standard deviation from this value, substrate porosity, as well as pore deviations from the round shape and average pore distance are given in Table 3.2.

The cross-section of the SiO_2/Si template (Fig. 3.7b) indicates that the pores are truncated cone-shaped. It also allows to determine the internal pore diameter (d), half-pore angle (θ) and silicon oxide thickness after etching (L). The schematically determined pore parameters are shown in Fig. 3.1a. According to the estimation, the pore parameters have the following values: $d \sim 160 \pm 10$ nm and $L \sim 440 \pm 10$ nm, with $\theta = 17$ °.

To determine the thickness of the silicon oxide layer and, correspondingly, the pore height L from the experimentally measured ellipsometry data $\Psi_e(\lambda)$, $\Delta_e(\lambda)$, the following model of the template structure was used [36, 37].

The optical parameters of the porous dielectric layers are calculated using an effective two-component Bruggeman environment [26]. In our case, the porosity of the template is sufficiently small and its effect on the refractive index is negligible. Therefore, we considered the possibility to use a simplified model of an isotropic dielectric layer of silicon dioxide on a silicon substrate. Fig. 3.8 shows the results of comparing the parameters of the computational model and the experimental data.

As can be seen, the theoretical model is in good agreement with experiment. The pore parameters of the SiO_2/Si template and the thickness of the oxide layer are correlated with the half-angle etching θ, given by the energy at the stage of

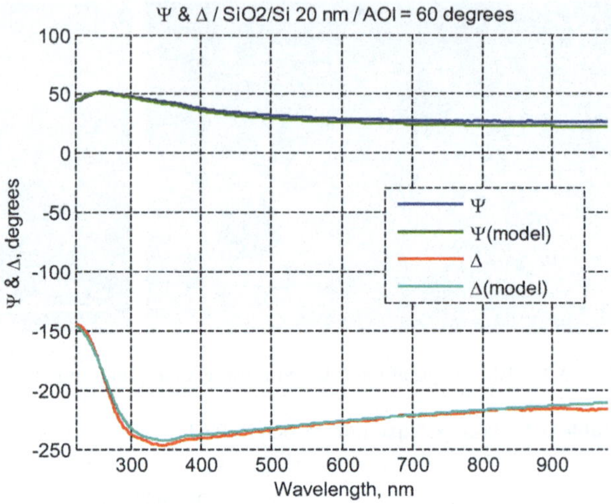

Fig. 3.8 The results of comparison of the physical model of the interaction of a polarized light beam with experimental data obtained from layered SiO_2/Si porous structure

irradiation with swift heavy ions [25]. Accordingly, to determine the height, outer and inner diameter of the pores, it is sufficient to know only one parameter in combination with the initial conditions L_{SiO2} and θ. The relationship between l and other a-SiO_2 pore parameters was determined by geometric Eqs. (3.2, 3.3, 3.4, 3.5 and 3.6), where θ was taken as a constant angle $\approx 17\,°$.

The results of comparing the thickness of a-SiO_2 using the ellipsometry method determined $L = 435$ nm. The values of the outer and inner diameters were calculated proceeding from formulas (1–2) and equaled 420 and 171 nm, respectively. The ellipsometry method provides information from a large sample area (the diameter of the observed region is from 10 to 200 μm, depending on the preliminary settings of the equipment). For example, on samples with a pore density of 5×10^7 cm^{-2}, ~50 pores enter the field of a polarized beam with a size of 200 μm in one measurement.

Thus, ellipsometry makes it possible to determine quickly and efficiently the averaged pore parameters in one measurement and to save considerable time during the research. In this case, L represents the average integral characteristic from a large area with an accuracy of up to 1 nm, which is due to the measurement of the relative (amplitude $\Psi(\lambda)$ and phase shift $\Delta(\lambda)$), rather than absolute values. One more advantage of ellipsometry compared with the SEM is an absence of necessity for the partial destruction of the sample (preparation of cross-section).

3.6 Conclusions

Ion-track SiO_2/Si templates were formed by irradiation with swift heavy ions and subsequent treatment in hydrofluoric acid. The study of oxide pores containing surface of silicon was made by using of scanning electron microscopy. Investigations of the dependences of characteristics of both individual pore and the template as a whole on the etching time and hydrofluoric acid concentration were carried out. The linear character of the dependence of pore parameters on the etching type and hydrofluoric acid concentration was determined. It was shown that standard deviation from the mean value of average area for each pore does not exceed 6% that indicates high quality of SiO_2/Si templates obtained by ion tracks chemical etching.

The technique of post-processing of obtained SEM images by using of ImageJ software package for obtaining and analyzing statistical data was described. The possibility of determining such template parameters as number of pores, their total area, average size and standard deviation, porosity, as well as deviations of pore shape from the circular and average distance between pores was demonstrated. Investigations of samples using the spectroscopic ellipsometry method made it possible to measure the thickness of SiO_2 layer, and the use of coupling equations allowed to calculate the upper and bottom diameters. The obtained as a result of SEM studies of the SiO_2 surface and SiO_2/Si template cross-section data were used for comparison with the results of restoring information about the pore parameters obtained by ellipsometry. Comparison of the different methods data shows a high degree of compliance that indicates the possibility of ellipsometry to rapid estimation of etched ion track parameters in silicon dioxide template.

Taking into account simplicity of SiO_2/Si templates preparation and small dispersion in the morphology as well as presence of both high-precision and express methods of templates quality control, using of ellipsometry method to control template parameters is an important asset for future integration of SiO_2/Si templates in different application fields.

Acknowledgements The authors acknowledge the support of the work in frames of H2020 – MSCA – RISE2017 – 778308 – SPINMULTIFILM Project, the Scientific-technical program 'Technology-SG' [project number 3.1.5.1], and Belarusian Foundation for Basic Research [project number Ф17M 005].

References

1. Xu D, Xu Y, Chen D, Guo G, Gui L, Tang Y (2000) Preparation of CdS single-crystal nanowires by electrochemically induced deposition. Adv Mater 12:520–522
2. Yakimchuk D et al (2018) Silver nanostructures evolution in porous SiO_2/p-Si matrices for wide wavelength surface-enhanced Raman scattering applications. MRS Commun 8:95–99
3. Sun L, Hao Y, Chien C-L, Searson PC (2005) Tuning the properties of magnetic nanowires. IBM J Res Dev 49:79–102
4. Demyanov S, Kaniukov E, Petrov A, Sivakov V (2014) Positive magnetoresistive effect in Si/SiO_2(Cu/Ni) nanostructures. Sensors Actuators A Phys 216:64–68
5. Fink D et al (2005) High energy ion beam irradiation of polymers for electronic applications. Nucl Instrum Methods Phys Res Sect B 236:11–20
6. Nishiyama H, Mizoshiri M, Hirata Y, Nishii J (2011) Hybrid microlens structures using femtosecond laser nonlinear lithography. IOP Conf Ser Mater Sci Eng 18:1–4

7. Fink D et al (2004) Etched ion tracks in silicon oxide and silicon oxynitride as charge injection or extraction channels for novel electronic structures. Nucl Instrum Methods Phys Res Sect B 218:355–361
8. Kaniukov EY, Shumskaya EE, Yakimchuk DV, Kozlovskiy AL, Ibragimova MA, Zdorovets MV (2017) Evolution of the polyethylene terephthalate track membranes parameters at the etching process. J Contemp Phys (Armenian Acad Sci) 52:155–160
9. Dallanora A et al (2008) Nanoporous SiO_2/Si thin layers produced by ion track etching: dependence on the ion energy and criterion for etchability. J Appl Phys 104:024307–1-024307-8
10. Giorgis F et al (2008) Porous silicon as efficient surface enhanced Raman scattering (SERS) substrate Appl. Surf Sci 254:7494–7497
11. Bandarenka H, Artsemyeva K, Redko S, Panarin A, Terekhov S, Bondarenko V (2013) Effect of swirl-like resistivity striations in n+−type Sb doped Si wafers on the properties of Ag/porous silicon SERS substrates. Phys Status Solidi 10:624–627
12. Kozlovskiy AL et al (2017) Comprehensive study of Ni Nanotubes for bioapplications: from synthesis to payloads attaching. J Nanomater 2017:1–9
13. Korolkov IV et al (2017) The effect of oxidation pretreatment of polymer template on the formation and catalytic activity of Au/PET membrane composites. Chem Pap 71:2353–2358
14. Kaniukov E et al (2017) Growth mechanisms of spatially separated copper dendrites in pores of a SiO 2 template. Philos Mag 6435:1–16
15. Kaniukov E, Kozlovsky A, Shlimas D, Yakimchuk D, Zdorovets M, Kadyrzhanov K (2016) Tunable synthesis of copper nanotubes. IOP Conf Ser Mater Sci Eng 110:012013
16. Bercu B, Enculescu I, Spohr R (2004) Copper tubes prepared by electroless deposition in ion track templates. Nucl Instrum Methods Phys Res Sect B 225:497–502
17. Sivakov V et al (2014) Silver nanostructures formation in porous Si/SiO_2 matrix. J Cryst Growth 400:21–26
18. Weeks RA (1956) Paramagnetic resonance of lattice defects in irradiated quartz. J Appl Phys 27:1376–1381
19. Holzenkampfeer E, Richter FW, Stuke J, Voget-Grote U (1979) Electron spin resonance and hoping conductivity of a-SiOx. J Non-Cryst Solids 32:327–338
20. Mazzoldi P et al (1991) N and Ar ionimplantation effects in SiO_2 films on Si sigleccrystal substrates. J Appl Phys 70:3528–3536
21. Trautmann C, Bouffard S, Spohr R (1996) Etching threshold for ion tracks in polyimide. Nucl Instrum Methods Phys Res Sect B 116:429–433
22. Awazu K, Kawazoe H (2003) Strained Si-O-Si bonds in amorphous SiO_2 materials: a family member of active centers in radio, photo, and chemical responses. J Appl Phys 94:6243–6262
23. Toulemonde M, Enault N, Fan JY, Studer F (1990) Does continuous trail of damage appear at the change in the electronic stopping power damage rate? J Appl Phys 68:1545–1549
24. Toulemonde M, Bouffard S, Studer F (1994) Swift heavy ions in insulating and conducting oxides: tracks and physical properties. Nucl Instrum Methods Phys Res Sect B 91:108–123
25. Kaniukov EY et al (2016) Tunable nanoporous silicon oxide templates by swift heavy ion tracks technology. Nanotechnology 27:115305
26. Fujiwara H (2007) Spectroscopic ellipsometry. Wiley, Chichester
27. Moreau WM (1988) Semiconductor lithography: principles, practices, and materials. Springer, New York, p 1988
28. Verhaverbeke S (1994) The etching mechanisms of SiO_2 in Hydrofluoric acid. J Electrochem Soc 141:2852–2857
29. Milanez SC, Varisco P, Moehlecke A, Fichtner PP, Papaléo RM, Eriksson J (2003) Processing of nano-holes and pores on SiO_2 thin films by MeV heavy ions. Nucl Instrum Methods Phys Res Sect B 206:486–489
30. Zhang WM et al (2008) Study of etched ion track profiles in silicon dioxide membrane. Radiat Meas 43:S627–S631
31. Bergamini F, Bianconi M, Cristiani S (2007) Wet and vapor etching of tracks produced in SiO_2 by Ti ion irradiation. Nucl Instrum Methods Phys Res Sect B 257:593–596

32. Rotaru CCS (2004) Stanescu SiO_2 sur silicium: comportement sous irradiation avec des ions lourds, Universite de Caen.
33. Knotter DM (2000) Etching mechanism of vitreous silicon dioxide in HF-based solutions. J Am Chem Soc 122:4345–4351
34. Yang Yu B, Elbuken C, Ren CL, Huissoon JP (2011) Image processing and classification algorithm for yeast cell morphology in a microfluidic chip. J Biomed Opt 16:066008
35. Bundyukova V, Yakimchuk D, Shumskaya E, Smirnov A, Yarmolich M, Kaniukov E (2019a) Post-processing of SiO2/Si ion-track template images for pores parameters analysis. Mater Today Proc 7:828–834. https://doi.org/10.1016/j.matpr.2018.12.081
36. Yakimchuk D, Bundyukova V, Smirnov A, Kaniukov E (2019) Express method of estimation of etched ion track parameters in silicon dioxide template. Phys Status Solidi B 256:1800316. https://doi.org/10.1002/pssb.201800316
37. Bundyukova V, Kaniukov E, Shumskaya A, Smirnov A, Kravchenko M, Yakimchuk D (2019b) Ellipsometry as an express method for determining the pore parameters of ion-track SiO2 templates on a silicon substrate. EPJ Web Conf 201:01001. https://doi.org/10.1051/epjconf/201920101001

Chapter 4
Optical and Electrical Properties of Ferric Chloride Doped Graphene

Marian Baah and Tommi Kaplas

Abstract In this work, we focus on synthesis, transfer and characterization of doped graphene. Growth of single layer graphene on copper foil was done by low-pressure chemical vapor deposition (CVD) technique. In the thermal CVD process, hydrogen was used as a process gas with methane as the carbon precursor. Graphene films were synthesized on Cu foil, which was later etched away. After the transfer, graphene films were doped with chemical dopant using ferric chloride ($FeCl_3$). Furthermore, graphene samples on silica substrates were characterized by Raman spectroscopy, spectrophotometry, scanning electron microscopy and four-point-probe technique. The doping process yielded $\sim$ 50% decrease in sheet resistance, and no changes in the transmittance or in the Raman spectra of the graphene sample.

Keywords Raman spectroscopy · Chemical doping · SEM · Four-point probe

4.1 Introduction

Graphene, a monoatomically thin layer of graphite was experimentally studied first time in 2004 by K.S. Novoselov and A.K. Geim [1, 2]. In their experimental study they used a mechanically cleaved, a few microns size graphene crystal. The tiny graphene flake was enough to demonstrate the existence of 2D graphite. However, it was obvious that for practical applications the size of the fabricated graphene films should be scaled up. In 2009, Li et al. and Reina et al. introduced first time chemical vapor deposition technique for graphene [3, 4]. In this technique, a multilayer graphene film was synthetized on a nickel foil [4]. Synthesis of a monolayer graphene on a nickel foil was rather challenging, and therefore, later on, the nickel substrate was replaced with copper. Because of low carbon solubility in copper, the synthesis of CVD graphene on copper results in a monolayer graphene

M. Baah (✉) · T. Kaplas
Institute of Photonics, University of Eastern Finland, Joensuu, Finland
e-mail: mariabaa@uef.fi

A. Maffucci, S. A. Maksimenko (eds.), *Fundamental and Applied Nano-Electromagnetics II*, NATO Science for Peace and Security Series B: Physics and Biophysics, https://doi.org/10.1007/978-94-024-1687-9_4

[5]. This was a major breakthrough, which provided a tool to synthetize almost arbitrary sized graphene films with very high quality [6, 7].

The CVD technique is almost a perfect tool to grow high crystalline, large area graphene films [8]. However, the major drawback of this technique is that graphene should be synthetized on a metal catalyst, like on copper. Followed by the synthesis, graphene is transferred on a dielectric substrate. It is well understandable that transfer of an atomically thin layer of carbon is a challenging task. However, it has become a rather standard operation in graphene related research.

The standard graphene transfer has following steps. At first graphene on a metallic catalyst is coated with a thin layer of polymer. Then metallic catalyst is removed by wet etching. Next, the graphene + polymer stack is placed on a dielectric substrate and finally the polymer film is removed by organic solvent. In graphene transfer process, it is worth noticing that graphene is conventionally grown on a copper foil and the copper foil is etched by ferric chloride ($FeCl_3$). Rather often ferric chloride is cleaned properly from graphene since otherwise the $FeCl_3$ dopes graphene [9]. However, for some applications like solar cells or touch screens it would be desirable that graphene would be doped.

Here we focus on doping graphene by $FeCl_3$. Strong doping levels affect graphene by Pauli blocking, which will affect the optical absorption of graphene. Moreover, because of the iron atoms, $FeCl_3$ is brownish color solvent, which may give a color to graphene. In this work, we measure transmittance and sheet resistance of our doped graphene to recover how the doping affects the color and the resistivity of a graphene film.

4.2 Theoretical Background

Transmittance measurements from graphene samples both doped and undoped allows one to estimate the electrical conductivity of these samples [10–12] both doped and undoped.

$$T = \left[1 + \frac{Z_0}{2}\,\sigma_{opt}\right]^{-2}, \tag{4.1}$$

where t is the thickness of the sample, $Z_o = 377\ \Omega$ is the free space impedance and σ_{op} is the optical conductivity of a transparent sample [11].

Literature has shown that every compound or chemical solution transmits, reflects or absorbs electromagnetic radiation at different wavelengths. Graphene is known to be a highly transparent film with transmittance value of approximately 97.7% over very wide spectral range, absorbing only a substantial percentage of incident radiation (2.3%) [13–15] even for such a transparent material. Increase of the number of graphene layers, substrate or doping may affect the optical transparency of graphene [16].

4.2.1 Doping

High electron mobility together with low optical absorption makes a monolayer graphene desirable material for transparent electrodes [17]. One main drawback in the performance of graphene on transparent electrodes in industry is its relatively high sheet resistance values attributed to the absence of free electrons [18]. A way to overcome this obstacle is the use of doping techniques.

Doping graphene will affect its optical properties. More precisely, if a photon that arrives into graphene, has energy less than two times Fermi energy, the photon is not absorbed. If photon's energy is higher than $2E_F$, it will be absorbed with 2.3% expectation.

When doping graphene, the Fermi level is shifted which increases the number of free electron carriers. In addition to electrical properties, doping affects also optical properties of graphene. The shift of the Fermi level in graphene (illustrated in (b) of Fig. 4.1) induces Pauli blocking which results from the inhibition of electron transitions once the arrival orbital states are already filled by electrons. This phenomenon leaves the graphene sample transparent to all photons with energies lower than $2E_F$ of the graphene [19].

From literature, there are three main doping mechanisms used for graphene; heteroatom doping, electrostatic field tuning and chemical doping. Heteroatom doping and chemical modification are able to open a small bandgap, and fine-tune the Fermi level of graphene simultaneously. Electrostatic field tuning however does not open as bandgap but is able to cause a change in the Fermi level of graphene [18, 20].

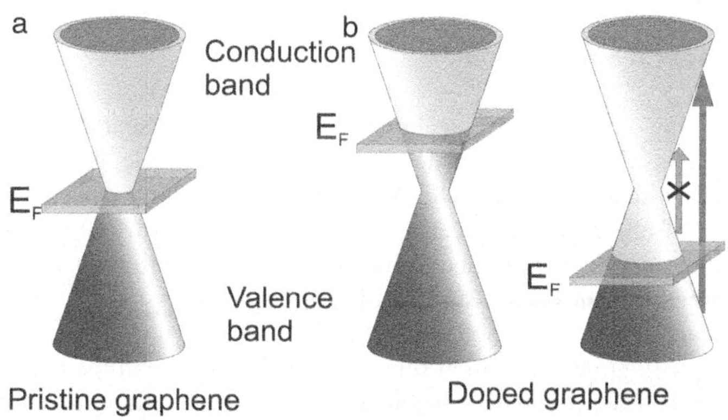

Fig. 4.1 The electronic band structure of pristine and doped graphene. (**a**) The band structure of pristine graphene indicating the valence and conduction bands, (**b**) the band structure of doped graphene

4.2.1.1 Heteroatom Doping

Heteroatom doping occurs when an acceptor atom or a donor atom is introduced into the graphene carbon atoms. For a hole acceptor element like boron it causes a p-type doping which in turn moves the Fermi level to the valence band of graphene. An electron donor element like Nitrogen causes n-type doping of graphene and moves the Fermi level of graphene to the conduction band at the same time opening a bandgap [18]. The heteroatom doping process occurs through the process of substitution, here some carbon atoms in the graphene are substituted with either Boron or Nitrogen atoms. A group of researchers at the Stanford University has been able to successfully dope graphene with Nitrogen atoms by electrothermally introducing NH_3 to graphene [18, 21, 22]; using a technique, they refer to as edge doping.

4.2.1.2 Electrostatic Field Tuning

This method of doping allows one to control the Fermi level as well as the electrical and magnetic properties of graphene without causing any more defects in the graphene film [18, 23]. By applying an electric field, perpendicular to graphene plane, the Fermi level is changed resulting in the increase of free charge carriers. By changing the direction of the applied electric field, one is able to tune the Fermi level of graphene from the conduction band to the valence band and vice versa [18]. As stated earlier, this form of doping does not open a bandgap in graphene but only changes the position of the Fermi level.

4.2.1.3 Chemical Modification

Chemical doping is the doping technique used in this research work. We say graphene film is chemically doped, when the film is made to interact with a chemical dopant such as $AuCl_3$ [18]. Two main processes may occur when using the technique of chemical modification to dope graphene, surface transfer doping or substitutional doping. In surface transfer doping, interaction between graphene and an adsorbed dopant on the surface of graphene occurs via the direct transfer of charge potential between graphene and the dopant or a series of redox reactions occurring between graphene, which serves as an electrode, and the dopant [18, 20, 24]. Substitutional doping occurs when there is the replacement of carbon atoms from graphene with impurity atoms from the dopant. Examples of chemical dopants used in chemical modification of graphene are, nitric acid (HNO_3), hydrogen sulphate (H_2SO_4), and toluene [25]. Here in this work our chemical dopant is $FeCl_3$.

4.2.2 Synthesis of Graphitic Carbon

Graphene synthesis ever since its discovery has gone through a series of techniques such as mechanical exfoliation, graphite oxide reduction, etc. One of most successful techniques, chemical vapor deposition (CVD) is a rapidly growing technique that allows the synthesis of high quality, structural defect free, and different number of layers of graphene [26–28]. CVD uses gaseous process and precursor gases, mostly hydrocarbons, for the synthesis of graphene and metal catalysts to facilitate the growth of crystalline graphene films.

Typically CVD is divided into homogenous and heterogeneous processes. The homogeneous process involves gas-to-gas reactions while heterogeneous are gas-to-solid reactions [29]. The homogeneous process results molecules that land on a substrate. Therefore, homogeneous processes are not conventionally used to grow highly crystalline materials. In graphene synthesis, when a graphene crystal is grown on a metallic catalyst substrate the dominating reaction is heterogeneous [29]. The growth of graphene using the CVD can take two forms, surface segregation (precipitation) process or surface adsorption process [26, 28]. Surface segregation occurs, when the carbon has high solubility with the metal catalyst used, whereas surface adsorption occurs when the solubility of the carbon in the metal is very low. Two main metal catalysts have been used over the years, copper (Cu) and nickel (Ni). Among these two metal catalysts, copper has a low dissolution of the carbon precursor as compared to nickel though the both have low processing costs as compared to other transition metals that can be used for CVD synthesis of graphene [28, 30] but their growth mechanisms are quite different from each other.

The growth mechanism for Cu uses the surface adsorption process. This is a much more advantageous process as one can easily synthesize monolayers of graphene [8, 26] as Cu has lower reactivity with carbon [26]. When using a Cu foil, the quality of graphene is dependent on its surface morphology and purity of the Cu foil as well as the process parameters for the growth of the graphene film [31].

The steps and defects on the surface of the chosen copper catalyst serves as nucleation sites that facilitates the growth of graphitic layers during the graphitization process. These nucleation sites are therefore highly saturated with carbon atoms [32] (Fig. 4.2d).

4.2.3 Graphene Transfer

These metal catalysts used to facilitate the growth of crystalline graphene layers cannot be left attached to these films as it would end up defeating the whole idea of minimizing or increasing the optoelectronic advantages of these graphene films. This then brings about the idea of transfer, where the films are transferred from the metal substrates to the desired substrates. There are two main types of transfer, dry and wet transfer [33].

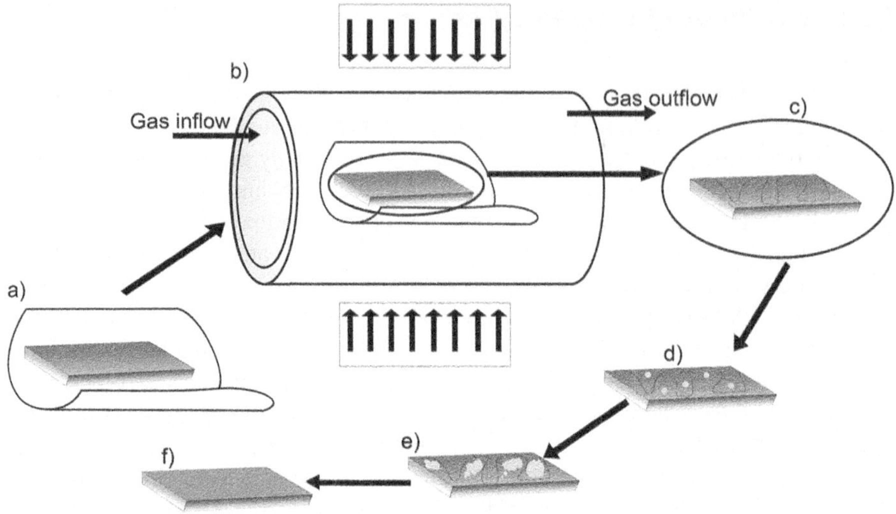

Fig. 4.2 Growth of graphene on copper catalyst. (**a**) Copper foil with a thin layer of copper oxide placed in CVD crucible, this layer can be removed by rinsing the copper foil in acetic acid or through the process of annealing [31]. (**b**) Crucible placed in CVD oven for graphitization process. (**c**) Insert showing the formation of copper grain boundaries; white spots represent nucleation centers for the growth of graphitic layer (**d**–**e**) and finally the coalescing of graphitic centers to form a layer of graphitic carbon on the Cu film (**f**)

4.2.3.1 Dry Transfer

Dry transfer technique involves the process of lamination where a supporting layer is deposited on the graphene on metal catalyst and a hot or cold lamination is applied. Lamination is done so the supporting layer, which is mostly polymer, is in direct contact with the graphitic layer. Once the polymer cools down, the graphene film is then transferred from polymer to the desired substrate [34].

4.2.3.2 Wet Transfer

Wet transfer technique involves the use of chemical etchants such as copper sulphate ($CuSO_4$) and hydrochloric acid (HCl) for nickel and ferric chloride ($FeCl_3$) for copper. In this technique, a supporting polymer is deposited on the surface of the graphitic layer on metal catalyst, baked to dry then the whole sample is immerse in the etching solution. Allowing this to sit until all the metal is etched away from the sample and the sample is lifted unto the target substrate. To remove this supporting polymer, the sample on substrate is fully immersed in an organic solvent and this dissolves all traces of the polymer leaving behind the graphitic film on the target or desired substrate [35–37].

4.3 Experimental Procedure

This section describes the experimental approach to which the samples of monolayer graphene was obtained from a large area graphene on copper target metal. Subsequent subsections describe the synthesis, transfer, doping and characterization techniques used.

4.3.1 *Synthesis of Graphene*

Large area graphene was grown on an 18 μm copper film substrate. A Cu foil of 10×10 cm^2 size was placed on a crucible and put into the CVD reactor. The chamber was pumped into vacuum for an hour with hydrogen gas flow of 10 sccm. After an hour of pumping, the CVD process began by chamber heating in H_2 atmosphere. The temperature of the chamber was made to rise from room temperature to 700 °C, at a heating rate of 20 °C/min and continued from 700 °C to 1000 °C at a heating rate of 10 °C/min. At 1000 °C, the Cu was annealed for 30 min to remove any remains of copper oxide and to increase the grain size of the Cu. CH_4 is then pumped into the chamber and a mixture of H_2 and CH_4 gases at 10 sccm and 20 sccm respectively is allowed to flow through the chamber for the next 30 min. After 30 min the chamber is pumped down to vacuum and cooled down in static H_2 atmosphere between 7.5 mbar and 10 mbar overnight.

4.3.2 *Graphene Transfer*

The graphene on copper is then spin-coated with an electron beam resist film AR-P 672 [38]. The resist consists of poly(methyl methacrylate) (PMMA). This polymer layer is to act as a support layer. The copper/graphene/PMMA is then dried for 5 min at 60 °C, this temperature is high enough to ensure that the polymer dries without turning the soft polymer into acryl glass [38]. The backside graphene, on the opposite side of the PMMA coated Cu substrate surface is then etched using the process of reactive ion etching (RIE) with oxygen plasma for 10 s/100 W/20 sccm. The device used for the back graphene etching is the Plasmalab 80. After the RIE process, the large samples is cut into 3 pieces and placed in a beaker containing diluted ferric chloride ($FeCl_3$) to etch away the Cu foil. The aqueous solution consisted of 90 g $FeCl_3$ and 600 ml deionized water. This solution is left overnight to ensure that all the Cu foil is completely removed leaving behind the graphene under the polymer layer. The film stack of graphene and polymer are transferred from the $FeCl_3$ solution into a beaker containing deionized water (Di-H_2O) for 60 mins. After 60 mins, the samples were again transferred from the first beaker of H_2O to another beaker of deionized H_2O and left for another 60 mins, this was done to ensure that most of the Cu foil remains and the etching solution is removed from the films. The films were then transferred unto the target substrate, fused silica (SiO_2) and left to air to dry.

All the samples were then dried again by baking on a hot plate for 20 mins at a temperature of 50 °C. This temperature is only high enough to evaporate all water residue from the sample without causing any bubbles or wrinkling to the sample. The thickness of the PMMA is then measured using the Dektak-150 Stylus Profilometer from Veeco and found to be ~200 nm. After measuring the thickness of the PMMA, all the samples were then placed in an organic solvent (acetone) and left overnight. The acetone is meant to remove all polymer remains from the samples. These samples were later rinsed in isopropanol, to remove all traces of acetone and any remaining polymer residue, for 5 min and dried with N_2 gas.

4.3.3 Graphene Characterization

The three samples produced from cleaned Cu foil were then characterized using Raman spectroscopy, scanning electron microscopy (SEM), four-point probe measurement technique and spectrophotometry.

4.3.3.1 Raman Spectroscopy

One spectroscopic technique that has proven to retrieve much information from graphene is Raman spectroscopy [39, 40]. This analytical tool has proven to identify and differentiate between sp^2 carbon films [39, 41]. The gapless nature of graphene allows all wavelengths incident on it to resonate, however, the Raman peak position from graphene may slightly change when using different excitation wavelengths [39]. Here, in this work we used 514 nm laser beam to probe graphene. Two main peaks characterize graphitic samples, the D ("disorder") and G ("graphite") peaks. The D peak is located near at 1350 cm^{-1} and is activated in the vicinity of a defect in the graphene lattice. The G peak is located around 1580 cm^{-1} and originates from the lateral stretching of sp^2 hybridized carbon atoms. For graphene and graphite, there appears a third peak, the 2D peak at the vicinity of 2700 cm^{-1},which originates from two phonon process [39]. Careful analysis of these features allows one to distinguish between a pristine and amorphous graphene film.

Raman spectra of all samples were measured using a Renishaw InVia Raman microscope, with the excitation wavelength of 514 nm connected and controlled by a computer interface. The low excitation wavelength was to ensure that there was no thermal effects involved, which might cause intrinsic effects in the samples [13].

4.3.3.2 Sheet Resistance Measurement

The four-point probe technique, which is designed specifically to measure the resistivity of semiconducting thin film samples. Especially, due to four measuring probes the contact resistance is suppressed.

The four-point-probe technique involves the use of a system consisting of four probes aligned collinearly and attached to a power supply. The two outer probes supply current to the samples while the two inner probes measure the voltage from the sample [42, 43].

The Signatone S – 1160 probestation with a direct current power source, was used for this work. It consists of four probes that are aligned collinear to each other and equally spaced with a probe spacing of 2 mm. It also consists of a power supply, three multimeters to measure the current (from outer probes) obtained from the sample, voltage (from the inner probes) supplied to the sample and voltage (2 V) from the power supply and a sample holder. A lever attached to the probestation allows the probes to be lowered unto the sample for measurement and raised from the sample after measurements.

Conventionally, graphene thin films are known to have sheet resistances ranging from 280 Ω/sq. to 1.2 kΩ/sq. [44]. These values are regarded as being too high for the effective performance of electronic devices. Through the process of doping, scientists have realized sheet resistances ranging from 600 Ω/sq. to as low as 58 Ω/sq. [44–46].

4.3.3.3 Scanning Electron Microscopy (SEM)

The graphene samples were visualized by the Leo (Zeiss) 1550 Gemini SEM for this work. This SEM device has a spatial resolution of 5 nm and the samples are required to be electrically conductive. However, our graphene samples on glass substrates were not coated with a conductive layer since they were conductive enough to be characterized as they were. The images produced from these samples gives information about the arrangement of the constituents of the samples and their topography[47, 48].

4.3.3.4 Spectrophotometric Measurements

Spectrophotometric measurements are made to obtain information concerning the optical properties of a specified sample. From these measurements, one may obtain the measure of transmittance, reflectance or absorbance of a sample. Especially, a monolayer graphene absorbs about 2.3% of light over very wide spectral range [44]. Therefore, by simple transmission spectrophotometry, the recognition of a monolayer graphene is rather straightforward.

For the spectrophotometry measurements, the Perkin Elmer Lambda 9 spectrophotometer was used. The device was calibrated by using a bare fuse silica substrate. The measurements were done at the wavelength range of 200–800 nm, which is a bit wider range in comparison to visual spectral range. An empty (uncoated) fuse silica substrate was used as a reference and the transmittance of all samples were measured. From these transmittance measurements the optical conductivity of the samples were estimated using Eq. 4.1.

4.3.4 Doping Graphene Samples

For graphene doping of the samples, they were each fully immersed in Di-H_2O: $FeCl_3$ (55: 45) solution for 60 s, at room temperature, then rinsed in deionized water (Di-H_2O) for another 60 s to remove most of the $FeCl_3$ molecules from the surface of graphene. Finally, samples were dried with nitrogen (N_2) gas.

We believe the chemical route to the chemical modification is as follows. Fe^+ ions are acceptor impurities since this compound dissociates in the form:

$$FeCl_3 \rightarrow Fe^{3+}_{(aq)} + 3Cl^-_{(aq)}, \tag{4.2}$$

therefore, the graphene layer becomes a p-type doped material after it has been immersed in the $FeCl_3$ solution.

By looking at graphene from the structural point of view, it is considered like a π-conjugated aromatic compound, then its doping can be likened to the doping process of other organic conducting polymers such as poly-(3-octylthiophene), doped with a metal chloride such as $AuCl_3$ [45]. The chemical reactions that takes place during doping with FeCl3 can then be analyzed as follows [45]:

$$2Graphene + 2FeCl_3 \rightarrow 2Graphene^+ + FeCl_2 - \left(Fe^I\right) + FeCl_4 \left(Fe^{III}\right), \tag{4.3}$$

$$3FeCl_2^- \leftrightarrow 2Fe^0 \downarrow + FeCl_4^- + 2Cl^- \tag{4.4}$$

This means that there is a reduction of Fe^{III} to Fe^I [45, 49], and this uncoordinated Fe^I disproportionate to form $FeCl_4^-$, Fe^I and Cl^-.

4.4 Results and Discussion

4.4.1 Results

In Fig. 4.3 the SEM images show graphene grown on copper, here we clearly see the copper grain boundaries indicated by the red arrow and the nucleation centers for growth of graphene, which we believe are the white spots.

The Raman spectra of the graphene on copper was also taken to ascertain the presence of graphitic film and to confirm the quality of the graphitic layer (graphene) on the copper substrate, this is represented in Fig. 4.4. Despite the wide baseline that originates from the copper substrate, the graphitic G and 2D peaks can be distinguished very clearly.

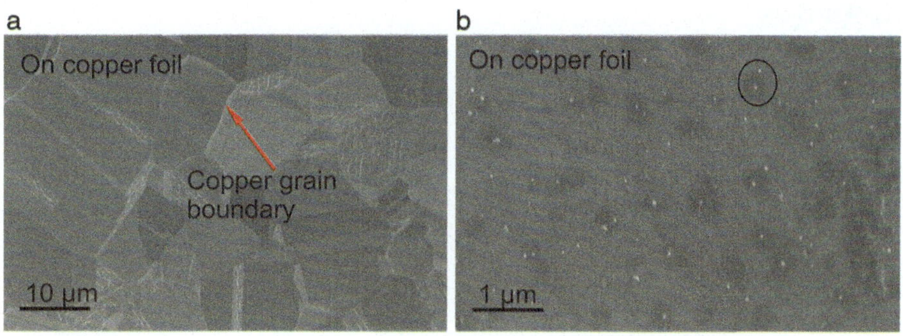

Fig. 4.3 Graphene on copper, (**a**) indicating copper grain boundaries and (**b**) a closer look reveals white spots (circle) are expected to be nucleation centers for the growth of graphene

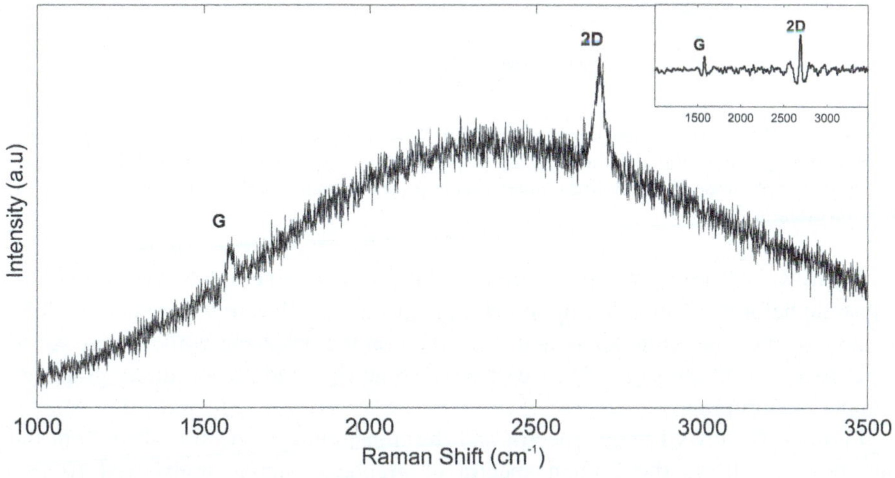

Fig. 4.4 Raman spectra of the graphene on Cu the insert is a representation of the detrended spectra, which clearly shows the two distinguishing peaks, G and 2D, for graphitic thin films and graphene

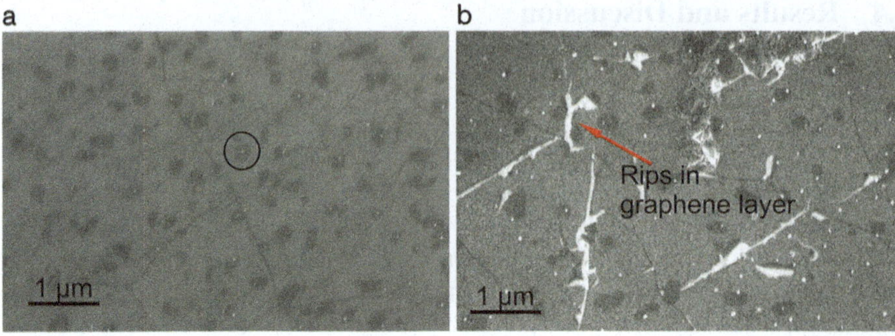

Fig. 4.5 (**a**) Undoped graphene transferred from copper and rinsed in acetone (ACE) and isopropanol (IPA), circle indicates the white spots and the dark regions around the spot shows regions of bilayer graphene formation. (**b**) Doped graphene with red arrow indicating rips in the doped graphene

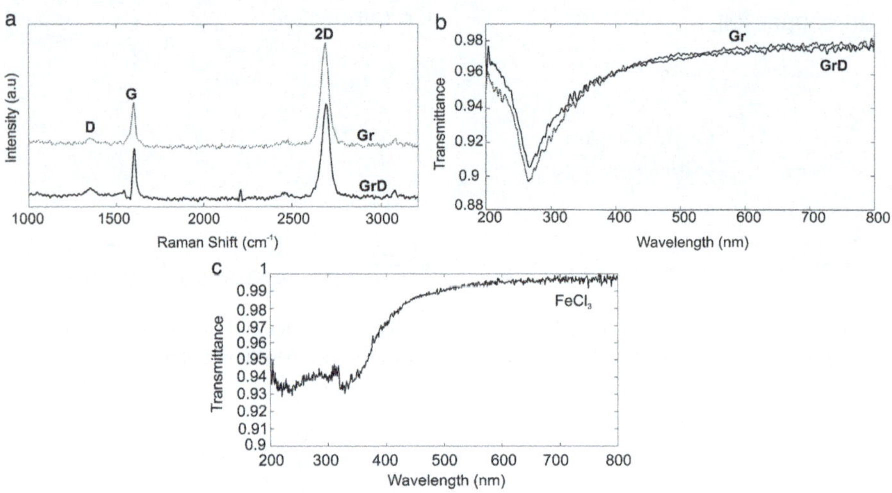

Fig. 4.6 (**a**) Raman spectra of graphene transferred from copper and rinsed in organic solvents, showing the spectra of the undoped (Gr) and doped (GrD) of the graphene samples. (**b**) Transmittance spectra of undoped and doped graphene sample. (**c**) Transmittance spectra of ferric chloride ($FeCl_3$) alone

Figure 4.5a, b are SEM representations of the transferred graphene on fused silica substrate before and after doping in $FeCl_3$, respectively. The circled area in Fig. 4.5a is indicative of the white spots that were seen on the graphene while it was yet to be transferred. In the Fig. 4.5b, we observe some rips (red arrow) in the graphene sample after doping.

Figure 4.6 shows Raman spectra and the transmittance spectra taken from the samples. (a) shows the Raman spectra of graphene that is transferred from a copper substrate unto a fused silica glass and later doped with $FeCl_3$. This shows dominating G and 2D peaks for the graphene sample but also negligible D peak that indicates very low number of defects in the graphite. The Fig. 4.6b shows the

Table 4.1 The summary of results obtained for undoped graphene (Gr) and $FeCl_3$ doped graphene (GrD). Raman measurements shows the peak positions and FWHM of the 2D, D and G peaks for graphene on copper, undoped graphene and doped graphene. Transmittance and four-point-probe measurements from which we obtained the DC conductivity (σ_{DC}) optical conductivity (σ_{OP}) and the ratio between σ_{DC} and σ_{OP} as well as sheet resistance

	Raman measurements						Transmittance and conductivity at 550 nm				
Sample	Peak position (cm^{-1})			FWHM (cm^{-1})			T	σ_{DC}	σ_{OP}	σ_{DC}/σ_{OP}	Sheet resistance Rs
	2D	G	D	2D	G	D	(%)	(S/m)	(S/m)		Ω/sq
On copper	2721	1591	–	35	24	–	–	–	–	–	–
Gr	2692	1596	1346	48	25	51	96.2	5192	2636	1.97	1284
GrD	2692	1597	1345	45	22	47	96.2	11,074	2636	4.2	602

transmittance measurements made with the doped and undoped graphene sample. Undoped and doped graphene are denoted by Gr and GrD, respectively.

By following the graphene doping routine we immersed an empty silica substrate to $FeCl_3$. Furthermore, we measure a transmittance from this sample (shown in Fig. 4.6c). The experiment shows that the transmittance of FeC13 immersed silica substrate is reduced especially at range from 200 nm to 350 nm.

Raman characterization and transmittance spectroscopy reveal that there are almost no differences in the optical properties of Gr and GrD. However, by four-point-probe technique, we observe that the sheet resistance (Rs)of graphene decreased more than 50%. More specifically, before doping the Rs of graphene was observed to be around 1200–1300 Ω/sq. By doping graphene in $FeCl_3$, the Rs value decreased close to 600 Ω/sq. The summary of the obtained results are presented in Table 4.1.

4.4.2 Discussion

Graphene has constant absorption over very wide spectral range [45]. Because of the constant absorbance, graphene is rather neutral and thus desirable material for applications that require a visually transparent electrode. However, ferric chloride consists of iron atoms. Due to these iron atoms the $FeCl_3$ solution has strong, brownish color. If graphene is doped with $FeCl_3$, the transparency of a graphene electrode may receive this color.

In Fig. 4.6c, we show a transmittance spectrum of bare silica substrate that is immersed in FeC13. In this sample the transmittance decreases at range from 200 nm to 350 nm from the original 100% to 96%. However, in our experiment we did not observe such kind of change of the transmittance of graphene film before and after doping. In fact, as it is shown in Fig. 4.6b the transparency of a graphene film remains exactly the same despite the sample is rinsed in $FeCl_3$ solution. The reason why the color of graphene remains unchanged is expected to originate from rather low concentration of $FeCl_3$ molecules on graphene. After immersing a graphene sample in $FeCl_3$ doping solution the graphene sample is thereafter rinsed in deionized water. The rinsing procedure removes most of the $FeCl_3$ molecules from the graphene surface and thus the color of graphene remains unchanged.

Despite, the color of a doped graphene film remains the same as it was before doping, the electrical conductivity is significantly improved. Specifically, in our experiment the DC conductivity of doped graphene is increased more than 50% in comparison to that of the undoped graphene.

Furthermore, we observed that during doping procedure graphene was partially damaged. The mechanism that caused the damage may originate from poor adhesion to the fused silica substrate. When graphene is immersed in $FeCl_3$ solution, which is then followed by rinsing it in water, poor adhesion may detach the graphene film causing fractures to the graphene film.

The poor adhesion between the graphene film and the substrate may originate from white spheres that are located in our sample (see Figs. 4.3b and 4.5). These spheres are grown on Cu and are likely, at least partially, under graphene thus preventing a proper contact between the graphene film and the substrate. The origin of these white spheres is assumed to be copper oxide on a copper substrate. Despite, the Cu substrates were treated in acetic acid to remove copper oxide, we believe some oxidation took place in the CVD chamber.

4.5 Conclusion

Graphene is a desired material for transparent electrodes. This is especially because in graphene it is possible to combine high transparency with low sheet resistance. However, before graphene is usable for transparent electrodes, it requires to be doped beforehand. Here in this work, we have shown that by doping graphene with $FeCl_3$ the sheet resistance of graphene can be reduced for more than 50%. This is an indicator that $FeCl_3$ works well as a dopant chemical for graphene. Moreover, we have shown that $FeCl_3$ does not have significant impact to the transparency of graphene, which is important for transparent electrodes applications.

More specifically, in our experiment we observed that the Rs value of an undoped graphene was reduced from 1.3 kΩ/sq. to 600 Ω/sq. by $FeCl_3$ doping. However, the chemical doping did not affect to the transmittance of graphene samples. Therefore, we believe that $FeCl_3$ could be very suitable chemical dopant for graphene. Especially, since conventionally graphene is transferred by using $FeCl_3$ as a copper etchant to remove copper substrate. Combining the transfer process and doping process could be therefore very appealing way to reduce steps that are required to produce doped graphene.

References

1. Novoselov KS, Geim AK, Morozov SV, Jiang DA, Zhang Y, Dubonos SV, Grigorieva IV, Firsov AA (2004) Electric field effect in atomically thin carbon films. Science 306(5696):666–669
2. Geim AK, Novoselov KS (2010) The rise of graphene. Nano Sci Technol A Coll Rev Nat J *1*:11–19

3. Li X, Cai W, An J, Kim S, Nah J, Yang D, Piner R, Velamakanni A, Jung I, Tutuc E, Banerjee SK (2009) Large-area synthesis of high-quality and uniform graphene films on copper foils. Science *324*(5932):1312–1314
4. Reina A, Jia X, Ho J, Nezich D, Son H, Bulovic V, Dresselhaus MS, Kong J (2009) Layer area, few-layer graphene films on arbitrary substrates by chemical vapor deposition. Nano Lett *9*(8):3087–3087
5. Gao L, Guest JR, Guisinger NP (2010) Epitaxial graphene on Cu (111). Nano Lett *10*(9):3512–3516
6. Polat EO, Balci O, Kakenov N, Uzlu HB, Kocabas C, Dahiya R (2015) Synthesis of large area graphene for high performance in flexible optoelectronic devices. Sci Rep 5:16744
7. Bae S, Kim H, Lee Y, Xu X, Park JS, Zheng Y, Balakrishnan J, Lei T, Kim HR, Song YI, Kim YJ (2010) Roll-to-roll production of 30-inch graphene films for transparent electrodes. Nat Nanotechnol 5(8):574
8. Mattevi C, Kim H, Chhowalla M (2011) A review of chemical vapour deposition of graphene on copper. J Mater Chem 21(10):3324–3334
9. Kaplas T, Bera A, Matikainen A, Pääkkönen P, Lipsanen H (2018) Transfer and patterning of chemical vapor deposited graphene by a multifunctional polymer film. Appl Phys Lett *112*(7):073107
10. Bao W, Wan J, Han X, Cai X, Zhu H, Kim D, Ma D, Xu Y, Munday JN, Drew HD, Fuhrer MS (2014) Approaching the limits of transparency and conductivity in graphitic materials through lithium intercalation. Nat Commun 5:4224
11. De S, Coleman JN (2010) Are there fundamental limitations on the sheet resistance and transmittance of thin graphene films? ACS Nano 4(5):2713–2720
12. Zhu SE, Yuan S, Janssen GCAM (2014) Optical transmittance of multilayer graphene. EPL (Europhys Lett) *108*(1):17007
13. Kaplas T, Svirko Y, Kuzhir P (2016) Synthesis of Pyrolytic carbon films on dielectric substrates. In: Fundamental and applied nano-electromagnetics \. Springer, Dordrecht, pp 227–238
14. Kaplas T (2013) Synthesis and optical characterization of Ultrathin carbon films. University of Eastern Finland, Joensuu
15. Suzuki S, Takamura M, Yamamoto H (2016) Transmission, reflection, and absorption spectroscopy of graphene microribbons in the terahertz region. Jpn J Appl Phys 55(6S1):06GF08
16. Nair RR, Blake P, Grigorenko AN, Novoselov KS, Booth TJ, Stauber T, Peres NM, Geim AK (2008) Fine structure constant defines visual transparency of graphene. Science 320(5881):1308–1308
17. Kim J, Cote LJ, Kim F, Huang J (2009) Visualizing graphene based sheets by fluorescence quenching microscopy. J Am Chem Soc 132(1):260–267
18. Guo B, Fang L, Zhang B, Gong JR (2011) Graphene doping: a review. Insciences J 1(2):80–89
19. Mak KF, Ju L, Wang F, Heinz TF (2012) Optical spectroscopy of graphene: from the far infrared to the ultraviolet. Solid State Commun 152(15):1341–1349
20. Berashevich J, Chakraborty T (2009) Tunable band gap and magnetic ordering by adsorption of molecules on graphene. Phys Rev B 80(3):033404
21. Wang X, Li X, Zhang L, Yoon Y, Weber PK, Wang H, Guo J, Dai H (2009) N-doping of graphene through electrothermal reactions with ammonia. Science 324(5928):768–771
22. Li X, Wang H, Robinson JT, Sanchez H, Diankov G, Dai H (2009) Simultaneous nitrogen doping and reduction of graphene oxide. J Am Chem Soc 131(43):15939–15944
23. Nigar S, Zhou Z, Wang H, Imtiaz M (2017) Modulating the electronic and magnetic properties of graphene. RSC Adv 7(81):51546–51580
24. Pinto H, Markevich A (2014) Electronic and electrochemical doping of graphene by surface adsorbates. Beilstein J Nanotechnol 5:1842
25. Lv R, Terrones M (2012) Towards new graphene materials: doped graphene sheets and nanoribbons. Mater Lett 78:209–218
26. Das S, Sudhagar P, Kang YS, Choi W (2014) Graphene synthesis and application for solar cells. J Mater Res 29(3):299–319

27. Kumar P, Singh AK, Hussain S, Hui KN, San Hui K, Eom J, Jung J, Singh J (2013) Graphene: synthesis, properties and application in transparent electronic devices. Rev Adv Sci Eng 2:1–21
28. Li X, Colombo L, Ruoff RS (2016) Synthesis of graphene films on copper foils by chemical vapor deposition. Adv Mater 28(29):6247–6252
29. Campbell SA (2001) The science and engineering of microelectronic fabrication, The Oxford series in electrical and computer engineering. Oxford University Press, New York
30. Muñoz R, Gómez-Aleixandre C (2013) Review of CVD synthesis of graphene. Chem Vap Depos 19(10–11–12):297–322
31. Kim SM, Hsu A, Lee YH, Dresselhaus M, Palacios T, Kim KK, Kong J (2013) The effect of copper pre-cleaning on graphene synthesis. Nanotechnology 24(36):365602
32. Aliofkhazraei M, Milne WI, Ozkan CS, Mitura S, Gervasoni JL, Ali N (2016) Graphene science handbook: size-dependent properties. CRC Press, Boca Raton
33. Chen Y, Gong XL, Gai JG (2016) Progress and challenges in transfer of large-area Graphene films. Advanced science 3(8):1500343
34. Fechine GJ, Martin-Fernandez I, Yiapanis G, Bentini R, Kulkarni ES, de Oliveira RVB, Hu X, Yarovsky I, Neto AHC, Oezyilmaz B (2015) Direct dry transfer of chemical vapor deposition graphene to polymeric substrates. Carbon 83:224–231
35. Liang X, Sperling BA, Calizo I, Cheng G, Hacker CA, Zhang Q, Obeng Y, Yan K, Peng H, Li Q, Zhu X (2011) Toward clean and crackless transfer of graphene. ACS Nano 5(11):9144–9153
36. De Arco LG, Zhang Y, Kumar A, Zhou C (2009) Synthesis, transfer, and devices of single-and few-layer graphene by chemical vapor deposition. IEEE Trans Nanotechnol 8(2):135–138
37. Regan W, Alem N, Alemán B, Geng B, Girit Ç, Maserati L, Wang F, Crommie M, Zettl A (2010) A direct transfer of layer-area graphene. Appl Phys Lett 96(11):113102
38. ALLRESIST, Innovation, Creativity, Customer-specific solutions (2014). https://www.allresist.com/wp.../2/.../allresist_produktinfos_ar-p630-670_englisch.pdf. Accessed 28 Aug 2018
39. Ferrari AC, Basko DM (2013) Raman spectroscopy as a versatile tool for studying the properties of graphene. Nat Nanotechnol 8(4):235
40. Binder J, Urban JM, Stepniewski R, Strupinski W, Wysmolek A (2015) In situ Raman spectroscopy of the graphene/water interface of a solution-gated field-effect transistor: electron–phonon coupling and spectroelectrochemistry. Nanotechnology 27(4):045704
41. Marshall CP, Edwards HG, Jehlicka J (2010) Understanding the application of Raman spectroscopy to the detection of traces of life. Astrobiology 10(2):229–243
42. Grigoryeva M (2014) Reducing contact resistance in graphene devices. http://urn.fi/URN:NBN:fi-fe2014061829334. Accessed 14 Mar 2019
43. Hasegawa S, Grey F (2002) Electronic transport at semiconductor surfaces—from point-contact transistor to micro-four-point probes. Surf Sci 500(1–3):84–104
44. Parvez K, Li R, Müllen K (2015) Graphene as transparent electrodes for solar cells. In: Nanocarbons for advanced energy conversion, vol 2. Wiley, Hoboken
45. Kim KK, Reina A, Shi Y, Park H, Li LJ, Lee YH, Kong J (2010) Enhancing the conductivity of transparent graphene films via doping. Nanotechnology 21(28):285205
46. Gunes F, Shin HJ, Biswas C, Han GH, Kim ES, Chae SJ, Choi JY, Lee YH (2010) Layer-by-layer doping of few-layer graphene film. ACS Nano 4(8):4595–4600
47. Hafner B (2007) Scanning electron microscopy primer. Characterization facility, University of Minnesota-Twin Cities, pp 1–29
48. Goldstein J (ed) (2012) Practical scanning electron microscopy: electron and ion microprobe analysis. ACS Publications, Springer
49. Abdou MS, Holdcroft S (1996) Gold-decorated poly (3-alkylthiophenes). Chem Mater 8(1):26–31

Chapter 5
Self-Organization of Plasmonic Nanostructures in Pores of Silica Template for SERS

Dzmitry Yakimchuk, Egor Kaniukov, Victoria Bundyukova, Sergey Demyanov, and Vladimir Sivakov

Abstract Plasmonic nanostructures can greatly impact on Surface-enhanced Raman spectroscopy (SERS) for molecular analysis of materials and biological objects. Here we consider the criteria for plasmon-active metals choice, and also give examples of Au, Ag, and Cu nanostructures obtained by the controlled self-assembly in limited pore volume of the SiO_2/Si template for SERS applications. Based on the price/durability/amplification ratio, silver nanostructures are considered in more detail. It is shown that, depending on the parameters of the SiO_2/Si template and the synthesis regimes, it is possible to obtain silver structures with the form of crystallites, "sunflowers" and dendrites. The effectiveness of the application for SERS of SiO_2(Ag)/Si systems with sunflower-like structures and dendrites is demonstrated. It is found that the dendrites can be used for SERS with the benefit of lasers of different wavelengths (473, 532, and 633 nm).

Keywords Ion track technology · Template synthesis · Plasmonic nanostructures · Dendrites · SERS

5.1 Introduction

Plasmonic is part of optics engaged in the study of the interaction of light with nanostructures (NSs) at the metal-dielectric interface [1–3]. The design of metallic NSs based on copper, silver and gold makes it possible to produce devices with nanometer-scale elements, which are the basis for Surface-enhanced Raman scattering (SERS) [4]. Using plasmonic surfaces as SERS-active substrates, it is

D. Yakimchuk (✉) · E. Kaniukov · V. Bundyukova · S. Demyanov
Scientific-Practical Materials Research Centre, NAS of Belarus, Minsk, Belarus
e-mail: dim2yakim@gmail.com

V. Sivakov
Leibniz Institute of Photonic Technology, Jena, Germany

A. Maffucci, S. A. Maksimenko (eds.), *Fundamental and Applied Nano-Electromagnetics II*, NATO Science for Peace and Security Series B: Physics and Biophysics, https://doi.org/10.1007/978-94-024-1687-9_5

possible to amplify the Raman scattering signal many times, which makes it possible to detect and study substances in ultra-low concentrations [5–7]. Moreover, the development of plasmonic NSs can influence the integration and miniaturization of electronics and photonic connectors, and also find application in the implementation of sensitive analytical instruments [8–10].

The enhancement of the Raman scattering signal is usually attributed to the electromagnetic phenomenon of surface plasmon resonance, which is accompanied by an intensification of the field in the immediate vicinity of the NS (at so-called "hot spots") in comparison with the field of the incident electromagnetic wave [11–13]. Another factor contributing to the increase in signal intensity is the chemical effect, which consists in charge transfer between the metal NSs and the analyte's molecule [5, 6, 14, 15], but this phenomenon has less effect, in comparison with the plasmonic effect. Since electromagnetic enhancement is the main mechanism contributing to amplification of Raman signal, today most of studies in this field focus on the development of new types of plasmonic NSs for ensuring a high amplification degree of the signal. The amplification degree depends on the type of metal NSs, as well as their size and shape [16–18]. A simple way to control the parameters of the NSs is using template synthesis [19–21].

In this chapter we will discuss the most used plasmonic NSs from copper, silver and gold, which can be obtained using template synthesis in SiO_2 pores on silicon. The methods of obtaining and controlling the shape of plasmonic NSs based on silver in SiO_2/Si pores will discuss in more detail, as well as the results of determining the effectiveness of their use for SERS investigations.

5.2 Experimental Procedures

Single-crystalline silicon wafers (12 ohm·cm of the *p*-type (100)) with the SiO_2 amorphous dielectric layer of thickness ~600–700 nm have been used as substrates for the swift heavy ion (SHI) irradiation. The irradiation has been carried out by ^{197}Au ions with energy 350 MeV and fluencies up to 10^8 cm^{-2} on the UNILAC linear accelerator (GSI Helmholtz Centre for Heavy Ion Research, Darmstadt, Germany). The etching of SHI-irradiated silicon oxide has been carried out in the fluoric acid (HF). The determination of the main dimensions of pores was realized on the samples cross-section with their subsequent investigation by the scanning electron microscopy (SEM) on the JEOL 7000F setup. The electroless wet-chemical electroless deposition (WCED) method has been used for the obtaining of plasmonic NSs in the pores of the SiO_2 layer from the aqueous solution of the copper sulphate 0.02 M, silver nitrate 0.02 M, gold chloride 0.02 M, and hydrofluoric acid 5 M. The process temperature was 35 °C. For this purpose the electrolyte has been thermostated in a water bath during 20 min before the deposition process and the deposition time has not been exceed more then 1 min.

Structural analysis of the silver NSs was carried out by field emission scanning electron microscopy (Carl Zeiss ULTRA 55, FE-SEM). EDX analysis was performed in the specimen chamber of an EDX coupled SEM BrukerXFlash® 6 30 SDD EDS Detector. The growth direction and crystal structure of the silver NSs were investigated using Electron Backscatter Diffraction (EBSD) (Bruker e-Flash^{HR} EBSD Detector + ARGUS™ FSE/BSE-Detector System). Combining FE-SEM and EBSD allows the determination of individual grain orientations, local textures, and point to point orientation correlations on bulk surfaces of polycrystalline materials [22, 23].

The standard dye molecule Nile blue was used to study the Raman signal enhancement induced by Ag NSs in different shape. To perform SERS measurements, 50 μl of an aqueous solution of Nile blue ($C_{20}H_{20}ClN_3O$) was dropped on the samples with NSs with an incubation time of 15 min, and subsequently rinsed in deionized water, and finally, dried in air. The Raman scattering measurements were performed at room temperature on a Nanofinder High End (Lotis TII, Belarus–Japan) confocal microscope-based setup. Lasers with wavelengths of 473, 532, and 633 nm were used as excitation source. The laser power at the samples' surface was kept at 50 μW and the typical single-spectrum acquisition time was 5 s. The backscattered light, not analyzed for its polarization, was dispersed using a spectrograph with 0.55 m focal lengths equipped with a 600 lines/mm single grating spectrometer with a spectral resolution around 3 cm^{-1} and detected by a cooled CCD (charge-coupled device) camera. The laser beam was focused through a "50×" microscope objective lens. The resulting spot diameter was about 1 μm.

5.3 Challenge the Problem

Raman spectroscopy is well-suited to the specifics of nanobiotechnology and has some advantages: ease of use together with a high resolution; possibility of non-destructive control; simplicity of sample preparation without any requirements for sizes and form; capability of water solutions investigations, etc. However, the Raman scattering has a very low intensity. This fact does not allow to investigate the ultra-small quantities of substances. Also, the use of Raman spectroscopy for investigations of the composition and structure of biological compounds often leads to the appearance of fluorescent background, which makes it difficult to interpret the data. The solution to these problems is to use the SERS [12, 24–26].

Based on that, nowadays, the development of new plasmon-active surfaces based on copper, silver and gold NSs is very important scientific and technological aspect.

The choice of the type of metal for the design of SERS-active substrates is usually limited to three basic criteria: the affordable price, the degree of signal amplification, and the durability. The most affordable are the surfaces based on copper NSs, followed by silver and gold, respectively. However, by the degree of signal amplification, copper NSs take the last place, yielding to gold and the most plasmon-active silver. The durability is determined by corrosion resistance, and in this case gold is a priority choice, because not oxidized. Silver and copper

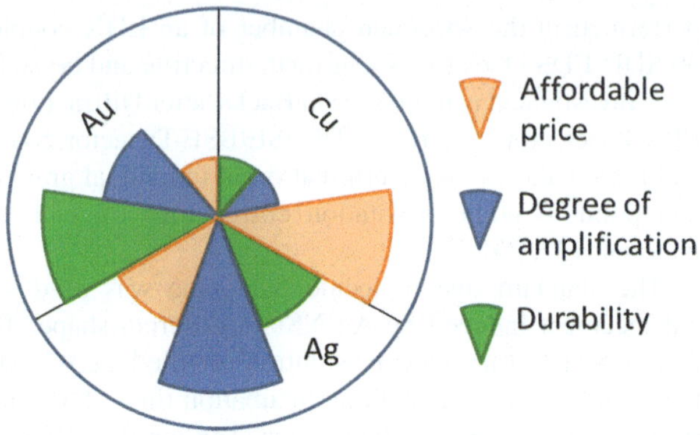

Fig. 5.1 Diagram of criteria for selecting/classification plasmon-active metals based on Au, Ag and Cu nanostructures. A larger area of the sector determines the advantage by the corresponding criterion of one metal over another

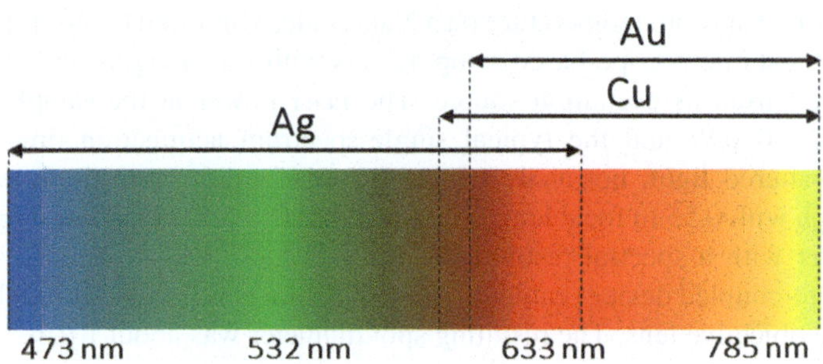

Fig. 5.2 Activity ranges of plasmonic NSs at visible and near-IR region

NSs degrade in time with forming on their surfaces oxide coatings that prevent the appearance of plasmon-active zones on the metal surface. Depending on the priority criteria, the plasmonic surface can be selected/classified using the diagram shown in Fig. 5.1.

Diagram in Fig. 5.1 shows that silver NSs are the most appropriated structures in the price/durability/amplification priorities. Also, this determines that the most publications are devoted to the Ag NSs as the basic elements of plasmon-active surfaces to the enhance of Raman signal.

The application of plasmonic surfaces for SERS purpose, their activity ranges are taken into account (distribution of the plasmon-active zones on the wavelength scale) [4]. The activity region of Au, Ag, and Cu NSs in the visible and near-IR wavelength range are shown on Fig. 5.2. For example, silver NSs exhibit plasmonic activity when exposed to radiation in the visible range from blue to red. The activity ranges of gold and copper structures are related more to red up to near-IR radiation. The exact position of the most intense plasmon-active regions is determined by the size and shape of the NSs [27]. For the convenience of choosing plasmonic NSs, the wavelengths of the most commonly used lasers for SERS are also indicated in Fig. 5.2.

Considering that signal amplification occurs on surface inhomogeneity [28], high-branched (dendritic) NSs are promising for SERS. The synthesis methods which are applied today provide the formation of dendrites based on plasmon-active metals lying in several layers on substrate surface [29–31], or forming 3D conglomerates [32–34]. The complexity of using of such systems, especially for studying of low concentrations, is connected with the penetration of the tested substance deep into/under the structure, limiting the access of the exciting radiation (laser) to the regions of analyte localization. Therefore, the signal from the analyte is not registered.

The size, shape and spatial distribution of metallic NSs can be controlled by using of template synthesis [35] method. Previously we showed that using ion-track templates, tubes [19, 36, 37], crystallites [38], dendrites [39, 40], and other forms of metallic NSs [41] can be obtaned. In this work, we present a simple method of design of plasmon-active Au, Ag, and Cu NSs, including dendritic-like forms, by using the controlled self-organization processes in the pores of the ion-track SiO_2/Si templates.

5.4 Choice of Optimal Plasmonic Nanostructures for SERS Applications

The pores in silicon oxide on single-crystalline silicon has been prepared by SHI-track technology, which includes chemical etching in HF. In our previous work [21] we have already demonstrated the creating SiO_2/Si templates with desirable characteristics. In preparating of the SiO_2 porous templates for plasmonic NSs creation, the nucleation process taking place exclusively on a silicon substrate has been taken into account. This fact has pre-determined the choice of the etching parameters ensuring the obtaining of the open pores. The formation of electroless deposition of plasmonic metal into the porous templates is schemaically represented in Fig. 5.3.

Figure 5.4 shows typical SEM images of Cu (Fig. 5.4a, b), Ag (Fig. 5.4c, d) and Au (Fig. 5.4 e, f) NSs, composed by individual nanoobjects with different shapes and sizes. It can be seen that the pore sizes of the template directly affect the shape of the structures and, in contrast to a number of existing works, do not repeat the geometry of the pores.

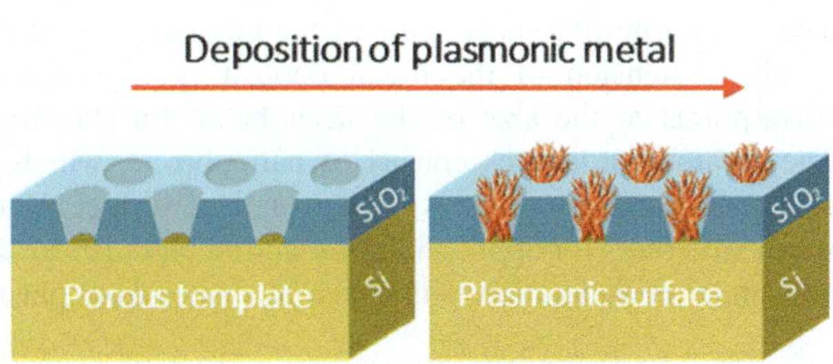

Fig. 5.3 Schematic representation of deposition of plasmonic silver nanostructures into porous Si/SiO_2 template

Fig. 5.4 SEM micrographs of self-organized copper (**a**, **b**), silver (**c**, **d**), and gold (**e**, **f**) NSs grown in porous SiO_2/Si template

In pores with diameters of 700–900 nm, sunflower-like NSs (Fig. 5.4a, c, e) formed, whereas at diameters of 500–700 nm, dendritic NSs (Fig. 5.4b, d, f) appeared. It is also shown that depending on the type of metal the crystallites on sunflower-like structures have different sizes. The copper structure consists of the smallest crystallites, in comparison to the increasing crystallite sizes by gold and silver NSs. Depending on the type of metal the geometry of dendritic-like structures can be varied. Gold (Fig. 5.4f) has a needle structure. Silver NSs (Fig. 5.4d) have mostly well-developed dendritic character with many branches, but rounded in comparison to gold NSs with more shapes geometry. Copper NSs (Fig. 5.4b) also have a dendrite-like shape with a large number of short branches.

The dimension of the metal deposit is $\sim$ 1 μm in pores of SiO_2/Si and corresponds to the size of the laser beam for the micro-Raman measurements. The obtained structures consist of nanoobjects, which creates conditions for the appearance of plasmon-active zones on their surface. Consequently, it is possible to select structures for application in required tasks for SERS by using of the series of NSs under consideration with different shapes from plasmonic metals.

Taking into account that Ag NSs is most preferable for SERS applications in terms of price/durability/amplification priorities, the silver NSs will be considered more detail in the next sections.

5.5 Silver Plasmonic Nanostructures Shape Evolution in Pores of SiO_2/p-Si Template

In this section, a more detailed study of the behavior of the shape of silver NSs, depending on the parameters of the SiO_2/Si template, is considered. For template synthesis of silver NSs we have used Si/SiO_2 wafers, which were irradiated by ^{197}Au ions with 350 MeV energy, fluence was 10^8 per cm^{-2}. In this work, it is demonstrated NSs, which are obtained using SiO_2/Si templates with different size of pores (D from ~200 nm to ~800 nm).

Recently, a simple WCED method [30, 42–44] was applied for silver NS formation using a hydrofluoric acid (HF) solution containing silver nitrate ($AgNO_3$). The procedure is suitable to synthesize unique silver dendritic NSs with stems, branches, and leaves. WCED in ionic metal (silver) HF solution is based on an electrochemical redox reaction in which both anodic and cathodic processes occur simultaneously at the silicon surface [45]. Ag NSs were formed on planar *p*-Si(100) and SHI irradiated SiO_2/Si surfaces by immersing samples in an aqueous solution of 0.02 M silver nitrate and 5 M HF in the volume ratio 1:1 at room temperature (20 ± 1 °C) for 30 s. Finally, the sample surfaces were rinsed several times in deionized water and dried at room temperature.

The formation of silver dendrites via galvanic deposition has been reported using bulk materials (e.g., silicon, aluminum, copper, zinc) as substrates [46–50]. In general, the above mentioned method normally involves the reduction of silver salts in the presence of potentiostatic or organic surfactants, at elevated temperature, or requiring the removal of template/ substrate to get pure products. Figure 5.5 shows

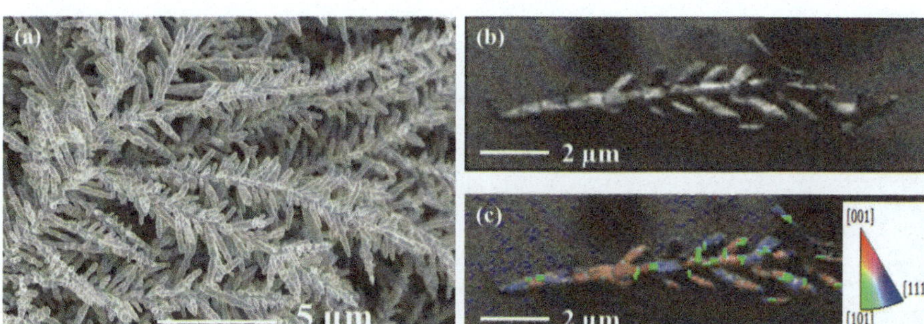

Fig. 5.5 (**a, b**) SEM micrographs of silver dendrite nanostructure on *n*-Si (100) grown for 30 s at atmospheric conditions; (**c**) EBSD analysis of single silver dendrite; colour coded inverse pole Figure representation [38]

typical Ag dendrite NSs synthesized by using a mixture of 5 M HF and $AgNO_3$ (0.02 M) in aqueous solution grown for 30 s on non-structured *n*-Si(100) substrate. The uniformity of a large number of such structures is illustrated in Fig. 5.5a. The dendrite structures with multilevel generations, having a long main trunk with short side branches (or even further sub-branches), all decorated by small leaves. Generally, the lengths of the trunks and (sub)-branches are several tens of micrometers and several micrometers, respectively. The diameters of all structures (trunk, branch, and sub-branches) have a similar size of few tens of nanometers.

The crystallinity character of the silver dendrites was investigated by EBSD of spatial resolution of a few 10 nm even over large probed areas allowing orientation analysis with respect to grain size (or single crystallinity), grain distribution, and silver orientations. Figure 5.5c gives evidence of single crystallinity of the silver dendrites over the full trunk length and branches with the presence of twins in the branching area.

The evolution of the NSs topology was studied to obtain a complete view of the Ag dendrite formation process and its growth mechanism. The pore diameter (top) was changed from approx. 200–800 nm by varying the etching time from 10 to 30 min in 5% HF solution, respectively. The silver formation is based on the galvanic displacement of Si by Ag^+ to Ag^0 reduction on the wafer surface [51]. The reaction proceeds in an aqueous solution of $AgNO_3$ and hydrofluoric acid. Ag^+ reduces onto the Si wafer surface by injecting holes into the Si valence band and oxidizing the surrounding lattice. The initial reduction of Ag^+ forms Ag^0 nanoparticles on the wafer surface. Further reduction of silver occurs on the newly formed nanoparticles, not the Si wafer, which becomes the active cathode by electron transfer from the underlying wafer. Figure 5.6 shows SEM images of silver nanostructures grown in 200 and 800 nm wide (top diameter) pores.

The size of the pores strongly influences the morphology of the silver NSs. In pores with a top diameter of 200 nm, the growth process yields single silver nanocrystals with little caps, while larger pores (800 nm in top diameter) lead to a totally different morphology with self-organized silver "sunflowers" (Fig. 5.6b), although the growth conditions were identical (0.02 M silver nitrate and 5 M HF,

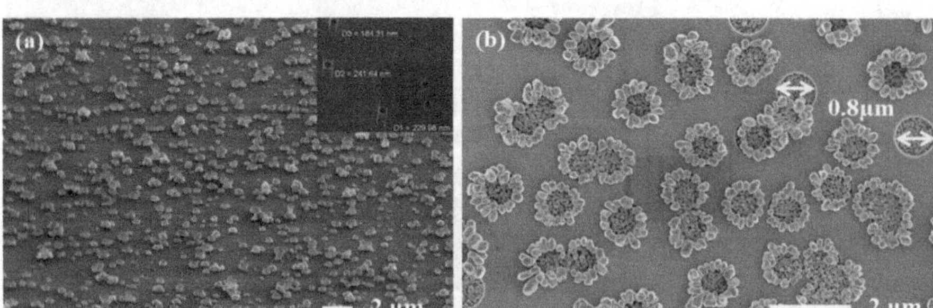

Fig. 5.6 SEM micrographs of silver nanocrystals selectively grown in pores with average top diameter (**a**) 200 nm (inset shows pores before Ag deposition) and (**b**) 800 nm [38]

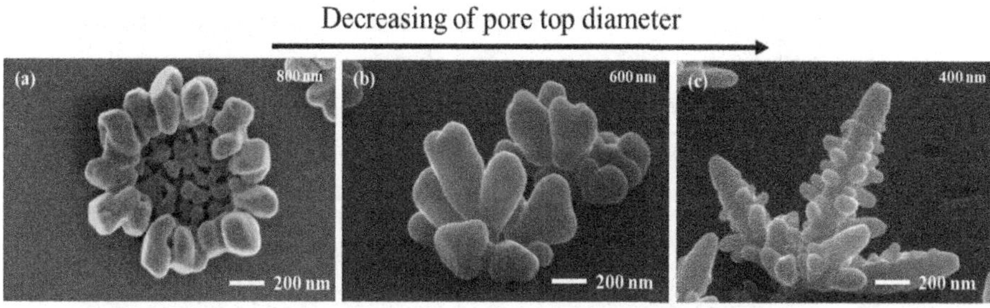

Fig. 5.7 SEM micrographs of silver nanostructures with (**a**) "sunflower", (**b**) "azalea" and (**c**) "corn"- like shapes [38]

room temperature; 30 s growth time). Tuning the parameters of the track etching process (e.g., concentration of silver cations in the etching solution, etchant temperature, etc.) produces quite different self-organized and reproducible shapes of "plasmonic nanogarden" objects" such as "sunflowers", "corns" or "azaleas" (Fig. 5.7). The growth mechanism of 3D silver NSs in porous matrix can be explained by the numerical calculations of particle trajectory and deposition using FEMLAB program (COMSOL) performed by Lee et al. [52] that corresponds that the column or 3D growth strongly depends on the particles concentration [52]. In our case it means that concentration of silver particles in pore less than on the surface between the pores that guided to the vertical growth of silver nanoparticles inside the pore, and the hierarchical structures grew on atop of pore where concentration of silver nanoparticles is higher and limited by the solution volume. The growth mechanism obviously depends strongly on the pore volume, pore formation (ion mass, energy, fluence, etching time) and silver deposition conditions (concentration, solution temperature and growth time) conditions.

5.6 Tuning of Silver Nanostructures Shape via Different Temperature Regimes

Considering the fact that, from a practical point of view, dendrites are the most suitable candidates for SERS, in this section we consider the effect of deposition temperature on the morphology of these structures. The self-organization of Ag NSs in the porous SiO_2/p-Si templates occur with a high organization degree of pore filling at all process temperatures that were investigated by SEM as presented in Fig. 5.8.

The morphology of silver deposited at room temperature is characterized by some variation in form of obtained Ag NSs, which is manifested in the fact that part of the crystallites, which form part of the metallic sediment in the pores have faceted edges and another part have smoothed edges (Fig. 5.8a, b). An increasing of silver deposition temperature to 25 °C leads to decreasing of the number of

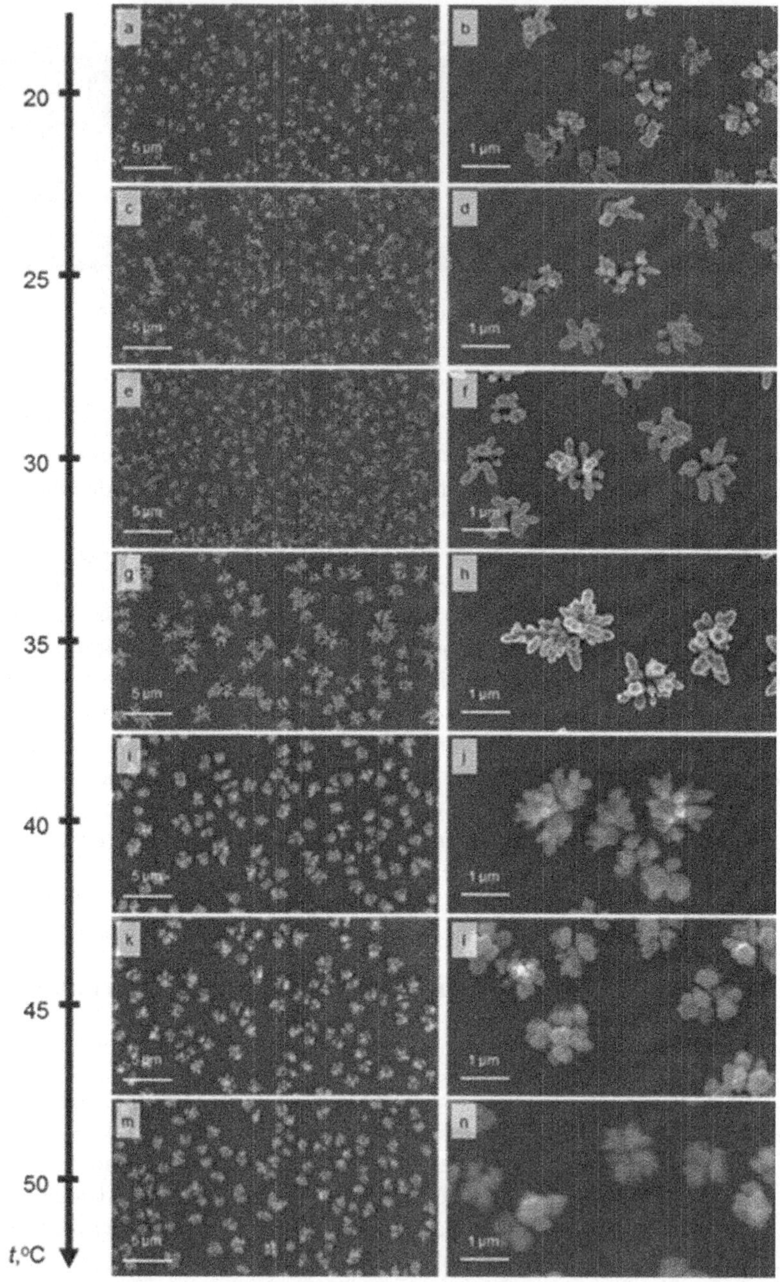

Fig. 5.8 Scanning electron images of silver NSs morphology evolution in porous SiO_2/p-Si templates depending on the silver deposition temperature variation [40]

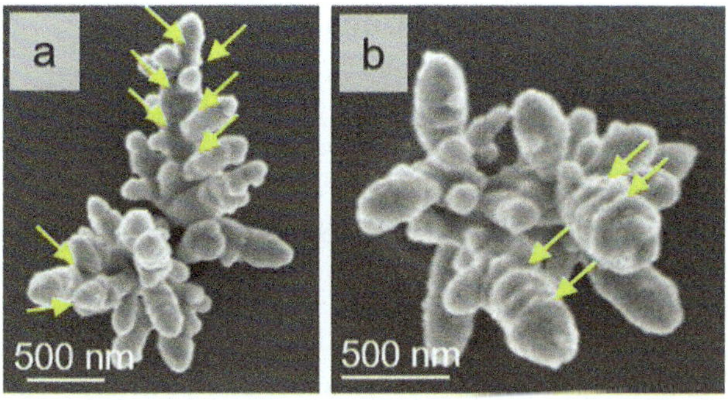

Fig. 5.9 SEM images of localized regions with "hot spots" on the surface of Ag NSs grown at 35 °C in the pores of the SiO_2/Si template: connecting parts of the branches (**a**) and the "bud-like" facets (**b**) of dendrite [40]

priority growth fronts with a simultaneous increasing of their lateral dimensions, while the structures are smoothed. Differences in the form of Ag NSs over the surface are minimal as shown in Fig. 5.8c, d. Further silver deposition temperature increasing to 30 °C does not lead any significant changes in the morphology of the silver sediment in comparison to 25 C, only insignificant increasing of its size was observed (Fig. 5.8e, f). An increasing of the temperature of the solution to 35 °C leads to a radical change in the shape of the nanostructures: it occurs an increasing of dimensions with filling of the central part of the pore, while on the branches of the Ag NSs appear branches of the second order and characteristic for "bud-like" facets of dendrites (Fig. 5.8g, h). Further silver growth temperature increasing to 40 °C leads to the destruction of the dendritic structure and formation of Ag NSs in different pores with different shapes: part of the metal deposit takes the form of faceted crystallites, and another part takes the form of the elongated smooth interconnected crystallites (Fig. 5.8k, l). Further increasing of temperature leads to decreasing of the number of smooth elongated (Fig. 5.8m, n) crystallites at 45 °C with their complete degeneracy at 50 °C (Fig. 5.8o, p).

The gaps between the dendrite branches (Fig. 5.9a) and the "bud-like" facets (Fig. 5.9b) are those nanoscale regions for which "hot spot" formation is characteristic [53–56]. Such structures are highly interesting for SERS applications. As it is known, the position of maximums of plasmonic structures extinction depends not only on their shape and size, but also on the polarization of the radiation, which induces plasmon oscillations [53, 57]. Accordingly, the same effects will be achieved, when the spatial orientation of the plasmonic structure changes: inclination, turning and etc. SEM images in Fig. 5.9 show that the resulting structures and their components (individual trunks and branches) are not characterized by strict spatial orientation relative to the template, but have a chaotic inclination and are not identical copies of each other. It is known that dendrites have a large set of "hot spots" [54, 56]. Thus, taking into account the influence of the polarization of the exciting radiation on the distribution of plasmons along the dendrite [53], it can be assumed that even a single silver structure has not one or several distinct bands of plasmon resonance. They are present in large numbers and are distributed throughout the long-wave interval corresponding to plasmon resonance bands of all possible Ag NSs [4].

5.7 SERS Applications

The SERS capabilities of the Ag NSs in porous SiO_2/Si templates can easily be performed by a simple model SERS experiments using the dye molecule NB as an agent of interest (analyte).

Figure 5.10 shows the SERS spectrum of NB molecules obtained from a sample with "sunflower" shaped NSs. Prior to the measurement, 150 μl of 10^{-5} M NB solution was dropped on the "sunflowers" surface for 15 min, afterward rinsed in water, and finally dried in the air. Under the excitation with a 532 nm laser light, the "sunflower" shaped silver nanostructures exhibit a high Raman signal as shown Fig. 5.10a. Besides the prominent band located at 521 cm^{-1} corresponded to crystalline silicon vibration, in the presented spectrum; all other bands are characteristic Raman modes for the NB molecule. As the "sunflowers" are relatively well-separated, it was possible to monitor the SERS signal of one individual "sunflower" by means of Raman mapping. The mapping was performed by scanning a 20×20 μm^2 square with step of 0.2 μm and monitoring the intensity of the 592 cm^{-1} band (phenoxazine ring mode) of NB (Fig. 5.10b).

The SERS spectrum obtained from incubating the "sunflower" shaped NSs with a diluted NB solution (10^{-5} M) exhibits a high Raman signal intensity giving clear evidence that such silver NSs grown in pores of etched SHI tracks are suitable for the Raman enhancement and are promising for broad exploitation for Raman analysis and imaging.

Taking into account the reasoning in Sect. 5.6 regarding the distribution of "hot spots" on dendritic structures, SERS measurements were made using lasers of different wavelengths: blue (473 nm), green (532 nm) and red (633 nm), as shown in Fig. 5.11.

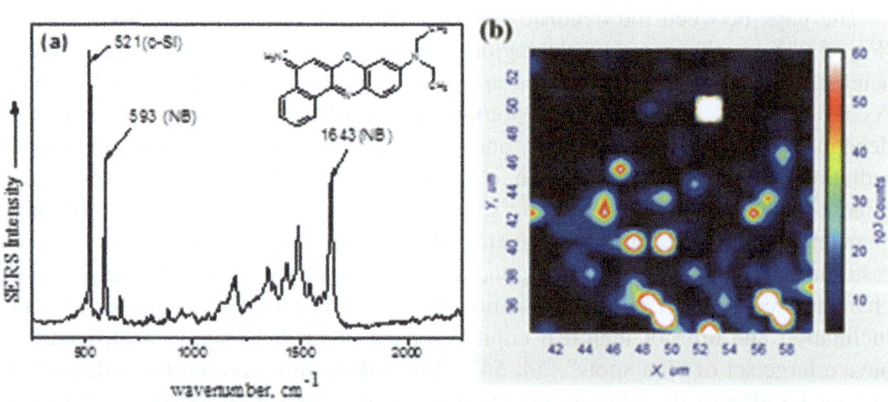

Fig. 5.10 (**a**) SERS spectrum of NB molecules (inset) obtained from a sample with "sunflower" shaped nanostructures; (**b**) 20×20 μm^2 map of individual "sunflower" shaped nanostructure showing the intensity of the Raman band at 592 cm^{-1} [38]

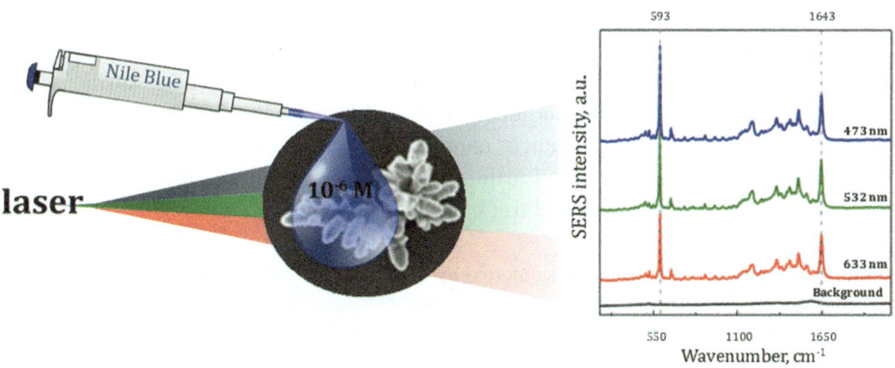

Fig. 5.11 Normalized SERS spectra of the NB molecule with concentration of 10^{-6} M, obtained on the surface of Ag NSs grown at 35 °C in the pores of the SiO_2/p-Si template, excited by using different excitation wavelengths (473, 532, and 633 nm), and the Raman spectrum of the substrate's background

The SERS spectra of NB with a concentration of 10^{-6} M have brightly expressed characteristic peaks at 592 cm^{-1} and 1643 cm^{-1} (Fig. 5.11), which correspond to the characteristic lines of NB [58, 59]. The intensity of the signal, when lasers with different wavelengths were used, indicates that the "hot spots" in the spectral region of 473–633 nm are uniformly distributed over the position and intensity of the plasmon resonance. A detailed study of this phenomenon with using of mathematical modeling with the interaction of an external electromagnetic wave with dendrites of different spatial orientations is planned to be carried out in the future work.

5.8 Conclusions

Simple NSs formation method of controlled self-organization in the pores of the ion-track SiO_2/Si template via WCED have been used for the formation of plasmon-active Au, Ag, and Cu NSs with various shapes. The shape of NSs depends on the templates pore size, chemical nature of metal and deposition conditions. The price/durability/signal amplification criteria diagram for selecting a plasmon-active NSs for realization of active SERS surfaces using in target application was proposed and analyzed. Based on our developed diagram the Ag NSs are the most appropriate choice for SERS applications.

The possibility of shape-controlled formation of silver NSs in SiO_2 template at various diameter of the pores and solution temperature was demonstrated as well as degree of amplification of a Raman signal of obtained silver NSs was studied. The "sunflower" shape silver NSs show SERS sensitivity allowing the detection of NB dye molecules. But from practical point of view the dendritic NSs is more interesting. Highly reproducible morphology and pore localized synthesis of the silver dendritic NSs in the SiO_2/Si templates allows to obtain the SERS-active substrates working in a broad spectral range (473–633 nm). These surfaces are expected to appeal to a wide audience in the nanotechnology related fields, like environmental science, food science, nanotoxicology, for studying the atmosphere composition in the study of near-Earth space, and etc. due to the high potential for in-depth studies of substances and track it behavior.

Acknowledgements This work was supported by the Scientific-technical 'program Technology-SG' (project number 3.1.5.1). V.S. is gratefully acknowledges the German Federal Ministry of Education and Research (BMBF) in frame of Baltic Sea Network "NanoPhoto" under Grant No. 01DS14017 and National Science Foundation of Germany (DFG) under Grant No. SI-1893/2-1 for the financial support.

References

1. Atwater HA (2007) The promise of plasmonics. ACM SIGDA Newsl 37:58–63
2. Ozbay E (2006) Plasmonics: merging photonics and electronics at nanoscale dimensions. Science 311:189–193
3. Maier SA, Atwater HA (2005) Plasmonics: localization and guiding of electromagnetic energy in metal/dielectric structures. J Appl Phys 98:011101
4. Sharma B, Frontiera RR, Henry A-I, Ringe E, Van Duyne RP (2012) SERS: materials, applications, and the future. Mater Today 15:16–25
5. Xia L et al (2014) Visualized method of chemical enhancement mechanism on SERS and TERS. J Raman Spectrosc 45:533–540
6. Zhao LL, Jensen L, Schatz GC (2006) Pyridine-Ag20 cluster: a model system for studying surface-enhanced Raman scattering. J Am Chem Soc 128:2911–2919
7. Arenas JF, Woolley MS, Tocón IL, Otero JC, Marcos JI (2000) Complete analysis of the surface-enhanced Raman scattering of pyrazine on the silver electrode on the basis of a resonant charge transfer mechanism involving three states. J Chem Phys 112:7669–7683
8. Barnes WL, Dereux A, Ebbesen TW (2003) Surface plasmon subwavelength optics. Nature 424:824–830
9. Haynes CL et al (2003) Nanoparticle optics: the importance of radiative dipole coupling in two-dimensional nanoparticle arrays †. J Phys Chem B 107:7337–7342
10. Maier SA (2004) Fundamentals and applications plasmonics: fundamentals and applications physics, vol 677. Springer, College Park, p 10
11. Santoro G et al (2014) Silver substrates for surface enhanced Raman scattering: correlation between nanostructure and Raman scattering enhancement. Appl Phys Lett 104:243107
12. Kneipp K (2007) Surface-enhanced raman scattering. Phys Today 60:40–46
13. Xia Y, Campbell DJ (2007) Plasmons: why should we care? J Chem Educ 84:91
14. Kneipp K (2016) Chemical contribution to SERS enhancement: an experimental study on a series of polymethine dyes on silver nanoaggregates. J Phys Chem C 120:21076–21081
15. Arenas JF, Woolley MS, Tocon IL, Otero JC, Marcos JI (2000) Complete analysis of the surface-enhanced Raman scattering of pyrazine on the silver electrode on the basis of a resonant charge transfer mechanism involving three states. J Chem Phys 112:7669–7683
16. Benz F et al (2016) SERS of individual nanoparticles on a mirror: size does matter, but so does shape. J Phys Chem Lett 7:2264–2269
17. Shaw CP, Fan M, Lane C, Barry G, Jirasek AI, Brolo AG (2013) Statistical correlation between SERS intensity and nanoparticle cluster size. J Phys Chem C 117:16596–16605

18. Wei H, Hossein Abtahi SM, Vikesland PJ (2015) Plasmonic colorimetric and SERS sensors for environmental analysis. Environ Sci Nano 2:120–135
19. Kaniukov E, Kozlovsky A, Shlimas D, Yakimchuk D, Zdorovets M, Kadyrzhanov K (2016) Tunable synthesis of copper nanotubes. IOP Conf Ser Mater Sci Eng 110:012013
20. Hillebrenner H, Buyukserin F, Stewart JD, Martin CR (2006) Template synthesized nanotubes for biomedical delivery applications. Nanomedicine (Lond) 1:39–50
21. Kaniukov EY et al (2016) Tunable nanoporous silicon oxide templates by swift heavy ion tracks technology. Nanotechnology 27:115305
22. Schwartz AJ, Kumar M, Adams BL (2000) Electron backscatter diffraction in materials science. Springer, Boston
23. Isabell TC, Dravid VP (1997) Resolution and sensitivity of electron backscattered diffraction in a cold field emission gun SEM. Ultramicroscopy 67:59–68
24. Uluok S, Guven B, Eksi H, Ustundag Z, Tamer U, Boyaci IH (2015) Designing multilayered nanoplatforms for SERS-based detection of genetically modified organisms. J Nanopart Res 17:43
25. Natan MJ (2006) Concluding Remarks: surface enhanced Raman scattering. Faraday Discuss 132:321
26. Moskovits M (1985) Surface-enhanced spectroscopy. Rev Mod Phys 57:783–826
27. Ding S-Y, You E-M, Tian Z-Q, Moskovits M (2017) Electromagnetic theories of surface-enhanced Raman spectroscopy. Chem Soc Rev 46:4042–4076
28. Shahbazyan TV, Stockman MI (2013) Plasmonics: theory and applications. Springer, Dordrecht
29. Brejna PR, Griffiths PR (2010) Electroless deposition of silver onto silicon as a method of preparation of reproducible surface-enhanced Raman spectroscopy substrates and tip-enhanced Raman spectroscopy tips. Appl Spectrosc 64:493–499
30. Qiu T, Wu XL, Mei YF, Chu PK, Siu GG (2005) Self-organized synthesis of silver dendritic nanostructures via an electroless metal deposition method. Appl Phys A Mater Sci Process 81:669–671
31. Qiu T, Wu XL, Shen JC, Xia Y, Shen PN, Chu PK (2008) Silver fractal networks for surface-enhanced Raman scattering substrates. Appl Surf Sci 254:5399–5402
32. Ye W, Shen C, Tian J, Wang C, Hui C, Gao H (2009) Controllable growth of silver nanostructures by a simple replacement reaction and their SERS studies. Solid State Sci 11:1088–1093
33. Sun X, Lin L, Li Z, Zhang Z, Feng J (2009) Novel Ag–Cu substrates for surface-enhanced Raman scattering. Mater Lett 63:2306–2308
34. Senthil Kumaran CK et al (2013) Preparation and characterization of Copper dendrite like structure by chemical method. Adv Mater Res 678:27–31
35. Kaniukov EY, Shumskaya EE, Yakimchuk DV, Kozlovskiy AL, Ibragimova MA, Zdorovets MV (2017) Evolution of the polyethylene terephthalate track membranes parameters at the etching process. J Contemp Phys (Armenian Acad Sci) 52:155–160
36. Kaniukov EY et al (2017) Electrochemically deposited copper nanotubes. J Surf Investig X-ray, Synchrotron Neutron Tech 11:270–275
37. Kozlovskiy AL et al (2017) Comprehensive study of Ni nanotubes for bioapplications: from synthesis to payloads attaching. J Nanomater 2017:1–9
38. Sivakov V et al (2014) Silver nanostructures formation in porous Si/SiO2 matrix. J Cryst Growth 400:21–26
39. Kaniukov E et al (2017) Growth mechanisms of spatially separated copper dendrites in pores of a SiO2 template. Philos Mag 97:2268–2283
40. Yakimchuk D et al (2018) Silver nanostructures evolution in porous SiO2/p-Si matrices for wide wavelength surface-enhanced Raman scattering applications. MRS Commun 8:95–99
41. Demyanov S, Kaniukov E, Petrov A, Sivakov V (2014) Positive magnetoresistive effect in Si/SiO2(Cu/Ni) nanostructures. Sensors Actuators A Phys 216:64–68
42. Peng K-Q, Yan Y-J, Gao S-P, Zhu J (2002) Synthesis of large-area Silicon nanowire arrays via self-assembling nanoelectrochemistry. Adv Mater 14:1164–1167

43. Sivakov V, Voigt F, Hoffmann B, Gerliz V, Christiansen S (2011) Wet – chemically etched silicon nanowire architectures: formation and properties. In: Hashim A (ed) Nanowires – fundamental research. InTech, Rijeka, pp 45–80
44. Sivakov V, Christiansen S (2012) Novel discovery of Silicon. J Nanoelectron Optoelectron 7:583–590
45. Gorostiza P, Kulandainathan MA, Díaz R, Sanz F, Allongue P, Morante JR (2000) Charge exchange processes during the open-circuit deposition of Nickel on Silicon from Fluoride solutions. J Electrochem Soc 147:1026–1030
46. Wang Z, Zhao Z, Qiu J (2008) A general strategy for synthesis of silver dendrites by galvanic displacement under hydrothermal conditions. J Phys Chem Solids 69:1296–1300
47. Shi F, Song Y, Niu J, Xia X, Wang Z, Zhang X (2006) Facile method to fabricate a large-scale superhydrophobic surface by galvanic cell reaction. Chem Mater 18:1365–1368
48. Yang Y, Meng G (2010) Ag dendritic nanostructures for rapid detection of polychlorinated biphenyls based on surface-enhanced Raman scattering effect. J Appl Phys 107:044315
49. Gutés A, Carraro C, Maboudian R (2010) Silver Dendrites from galvanic displacement on commercial Aluminum foil as an effective SERS substrate. J Am Chem Soc 132:1476–1477
50. He L, Lin M, Li H, Kim N-J (2009) Surface-enhanced Raman spectroscopy coupled with dendritic silver nanosubstrate for detection of restricted antibiotics. J Raman Spectrosc 41: 739–744
51. Peng KQ et al (2006) Fabrication of single-crystalline Silicon nanowires by scratching a Silicon surface with catalytic metal particles. Adv Funct Mater 16:387–394
52. Lee H et al (2011) Three-dimensional assembly of nanoparticles from charged aerosols. Nano Lett 11:119–124
53. Feng C, Zhao Y, Jiang Y (2015) Silver nano-dendritic crystal film: a rapid dehydration SERS substrate of totally new concept. RSC Adv 5:4578–4585
54. Zhao B et al (2016) Silver dendrites decorated filter membrane as highly sensitive and reproducible three dimensional surface enhanced Raman scattering substrates. Appl Surf Sci 387:431–436
55. Qiu T et al (2009) Hot spots in highly Raman-enhancing silver nano-dendrites. J Phys D Appl Phys 42:175403
56. Yin HJ et al (2015) Ag@Au core-shell dendrites: a stable, reusable and sensitive surface enhanced Raman scattering substrate. Sci Rep 5:1–9
57. Yang ZL et al (2010) FDTD for plasmonics: applications in enhanced Raman spectroscopy. Chin Sci Bull 55:2635–2642
58. Li WH et al (2013) Shape and size control, plasmonic properties, and use as SERS probes and photothermal agents. J Am Chem Soc 135:7098
59. Etchegoin PG, Le Ru EC (2010) Resolving single molecules in surface-enhanced Raman scattering within the inhomogeneous broadening of Raman peaks. Anal Chem 82:2888–2892

Chapter 6
Polymer Nanocomposites with Hybrid Fillers as Materials with Controllable Electrodynamic Characteristics for Microwave Devices

Ludmila Y. Matzui, Olena S. Yakovenko, Ludmila L. Vovchenko, Oleg V. Lozitsky, Viktor V. Oliynyk, and Volodymyr V. Zagorodnii

Abstract The results of investigation of microwave electromagnetic properties in wide frequency range (1–67 GHz) of the hybrid polymer composites based on combined introduction of ferrite and MWCNTs nanoparticles are presented. The surge in 3–5 orders of the conductivity value has been founded in MWCNTs/Fe_3O_4 ($BaFe_{12}O_{19}$)/epoxy composites with 2 and 5 wt. % of MWCNTs as compared with MWCNTs/epoxy composites, and these changes are more considerable for lower content of MWCNTs in the epoxy matrix. The addition of 30 wt. % of Fe_3O_4 ($BaFe_{12}O_{19}$) nanoparticles to MWCNTs/epoxy composites leads to the significant increase of the complex permittivity values and to the frequency shift of relaxation maximums. The natural ferromagnetic resonance was revealed as the main mechanism of magnetic loss in MWCNTs/Fe_3O_4/epoxy composites at frequencies up to 10–12 GHz while the eddy current induced mechanism becomes the dominant in magnetic loss at the further frequency increase. The natural ferromagnetic resonance determines the character of magnetic loss in $BaFe_{12}O_{19}$-filled composites in the whole investigated frequency range.

The integration of nanotubes and ferrite admixture of Fe_3O_4 or $BaFe_{12}O_{19}$ in MWCNT/30 wt. % Fe_3O_4 (or 30 wt. % $BaFe_{12}O_{19}$) ternary composites provides synergetic effects and complementary behavior (e.g., various polarization relaxations as additional loss mechanisms), which greatly contribute to the microwave absorption properties for both studied ternary composites. The absorption bandwidth reaches 13–17 GHz for thin (0.6–0.8 mm) composite samples with 2 wt. % of nanotubes content. The increase of MWCNTs content and the use of two

L. Y. Matzui (✉) · O. S. Yakovenko · L. L. Vovchenko · O. V. Lozitsky
Department of Physics, Taras Shevchenko National University of Kyiv, Kyiv, Ukraine
e-mail: matzui@univ.kiev.ua

V. V. Oliynyk · V. V. Zagorodnii
Department of Radio Physics, Electronics and Computer Systems, Taras Shevhenko National University of Kyiv, Kyiv, Ukraine

A. Maffucci, S. A. Maksimenko (eds.), *Fundamental and Applied Nano-Electromagnetics II*, NATO Science for Peace and Security Series B: Physics and Biophysics, https://doi.org/10.1007/978-94-024-1687-9_6

component fillers lead to decrease of the effective reflection loss in the frequency range above 20 GHz. The maximum reflection loss reaches 49 dB for Fe_3O_4- and 39 dB for $BaFe_{12}O_{19}$-filled composites of 1.1–1.2 mm thickness while the absorption bandwidth does not exceed 5 GHz.

Keywords Polymer composite · Nanocarbon · Hybrid filler · Ferrite · microwave shielding

6.1 Introduction

Currently there is ongoing tremendous development in the radio frequency (RF) and microwave (MW) electronics: mobile communications, high-speed electronic switching components and circuits, radars, navigation, etc. [1]. To safeguard the device operation, the nearby electronic equipment operating in the RF and MW frequency bands should be properly shielded. In particular, this can be achieved by coating the case of the device with some shielding material. Many of high frequency electronic devices are being operated in X-band (8–12 GHz) of electromagnetic spectrum The development of prospective communication standards requires high-speed data transmission and, correspondingly, new effective materials operating in the frequency range of 20–100 GHz. Furthermore, the development of soft and flexible electronic devices, such as foldable phones, electronic paper displays and wearable devices requires thin and flexible electromagnetic radiation (EMR) shielding films. All indications are that EMR shielding materials for the next-generation electronics will need seamlessly networked hetero-nanostructures the composition of which is the subject of many modern researches. Structured organic–inorganic composites have been extensively studied because they combine the advantages of inorganic materials (electric and magnetic properties, mechanical strength, thermal stability) and organic polymers (dielectric properties, processability, flexibility, and ductility). These composites combine characteristics of the individual constituent that can enhance electromagnetic absorption efficiency. Electromagnetic (EM) absorbing materials are mostly expected to have a broad bandwidth, high reflection loss (RL), and small thickness or light weight. Bandwidth is one of the most important parameters characterizing an EMR absorbing material in terms of real applications.

Nanocarbon materials and polymer nanocomposites (NC) have many advantages for potential applications as wideband EMR shielding materials: high shielding efficiency (SE), light weight, flexibility, high conductivity, and low contact resistance values. However, they have a number of shortcomings as microwave absorbers. First, a large volume of filler is needed to provide strong MW absorption, i.e. high values (often more than 40 dB) of EMR SE required for electronic devices. High content of carbon fillers would deteriorate the dispersion of the filler and thus would greatly affect the mechanical properties of the composite. Moreover, a conductive network produced by increased filler content will keep the microwave EM field outside the main composite volume and lead to limited microwave absorption. What is worse, the increased permittivity resulted from the increased filler content is accompanied by unwanted restriction of the effective bandwidth (EB). Meanwhile, it is rather difficult to obtain higher values of EMR SE for polymer/conductive filler composites, especially when the sole type of conductive filler is used. The addition of another inorganic filler is one of effective ways to enhance EMR absorption. The use of a mixture of closely packed nanocarbon and WS_2, MoS_2, or $BaTiO_3$ nanoparticles [2–5] supplies improved dielectric properties of the composite.

The mix of nanocarbon fillers with magnetic particles can achieve the enhancement of carbon-filled composites absorptive properties. Such magnetic admixtures can be ferrite or metal (alloy) particles. As compared with metallic (alloy) particles, ferrite particles offer important advantages of high ferromagnetic resonance frequency, high electric resistivity, low density, and good chemical stability. The addition of magnetic particles such as Fe_2O_3 or Fe_3O_4 [6, 7], $MnFe_2O_4$ [8], ZnO hollow spheres or nanoparticles [9], Co_3O_4 [10], MnZn-ferrite [11] could enhance the EMR SE and decrease the cost of NC. For instance, Li et al. [12] coated Fe_3O_4 on CNTs, and the specific RL_{min} of 28.7 dB and effective bandwidth of 8.3 GHz was achieved. The excellent absorbing properties were obtained in [13] for microstructured filler of interconnected network of graphene and carbon nanotubes with magnetic Fe_3O_4 particles. Authors demonstrate the superb potential for application of such structure as EMR absorbing material. The maximum RL of CNT film-Fe_3O_4-rGO-polydimethylsiloxane (PDMS) composite with 4 layers is 50.5 dB and absorption bandwidth at 10 dB is 5.7 GHz (the thickness of that material is 1.42 mm).

The most suitable candidates for EM absorbing materials in this frequency range are composites based on barium hexaferrite $BaFe_{12}O_{19}$, since it is characterized by excellent magnetic properties such as significantly high magnetocrystalline anisotropy and high ferrimagnetic–paramagnetic transition temperature [14, 15]. Hexagonal ferrites demonstrate large anisotropy, besides as permanent magnetic materials and perpendicular recording materials.

However, the practical application of the above-mentioned materials is severely limited due to some impeding factors. The analysis of publications [16–19] shows that microwave absorption materials performance depends on many factors: the type of nanocarbon and inorganic fillers, their morphology and structure, the processing method, the choice of the polymer matrix, the filler distribution inside the matrix, the interfacial interaction between fillers and matrix, etc. Ferrites demonstrate considerable magnetic loss and absorbing capability but are limited to applications due to their high density and processing difficulties. The applications of nanocarbons as fillers for EMR absorbers are hindered by missing magnetic loss and narrow absorption bandwidth. To overcome these complications, hybrid materials are developed that combine several materials to offset respective limitation of the materials retaining enhanced shielding properties. Certain success has been reached in enhanced absorbing properties of polymer composites with combined filler [16–19]. However, the overpowering majority of research reports refer to frequency range under 20 GHz where hybrid nanocarbon filler chemically modified by magnetic particles was used. So, the problems include finding the correlation between structure, morphology and phase composition of constituents and the properties of microwave absorption materials (MAM) in microwave frequency range. That will permit to determine the optimal composition of constituents and to synthesize materials and structures with required tunable EMR absorbing properties.

In present work the experimental results on EM properties in wide frequency range (1–67 GHz) with microwave absorption capability of the hybrid polymer composites based on combined application of ferrites and MWCNTs nanoparticles as fillers are shown.

6.2 Materials and Methods

6.2.1 Experimental Details

Structural and morphological peculiarities of NCs were studied by the optical microscopy ("Mikmed-1" with ETREK DCM-510 attachment) and the electron microscopy (JEOL JSM-6490LV; Mira3 Tescan).

The study of the electric resistance of the nanocarbon-polymer NC at the room temperature was performed by a four- or two-probe methods.

A coaxial method was used to determine the electromagnetic parameters of the investigated composites by using an Agilent N5244A vector network analyzer (Keysight, USA) in the frequency range of 1–67 GHz. Samples for the investigations were prepared in the toroid shape. The toroids tightly fit into the coaxial measurement cell. The cell (Fig. 6.1) has inner diameter of 0.8 mm and outer diameter of 1.85 mm. Full two-port calibration was initially performed on the test setup in order to remove errors due to the directivity, source match, load match, isolation and frequency response in both the forward and reverse measurements. The complex permittivity (ε_r) was determined from the scattering parameters using the Nicolson-Ross models.

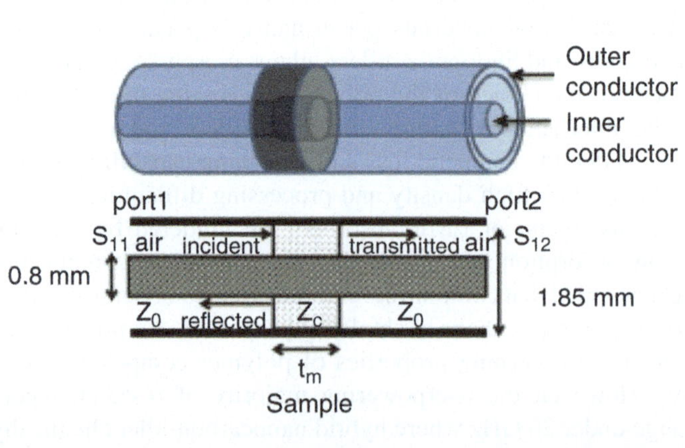

Fig. 6.1 The measurement scheme of dielectric spectra. Coaxial line and sample

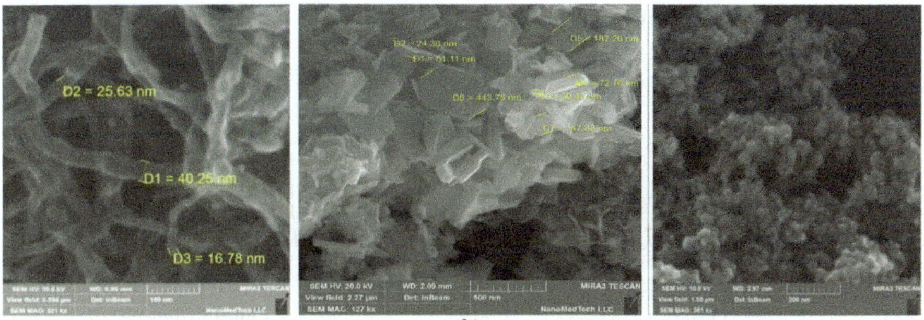

a) *b)* *c)*

Fig. 6.2 SEM images of (**a**) MWCNTs, (**b**) $BaFe_{12}O_{19}$, (**c**) Fe_3O_4 particles

Table 6.1 Morphology of the particles of fillers of polymer composites

	MWCNTs	$BaFe_{12}O_{19}$	Fe_3O_4
Shape	Cylinders	Plates	Spheres
Length	10–30 μm	–	–
Diameter	10–30 nm	150–450 nm	10–20 nm
Thickness	–	20–50 nm	–
Aspect ratio	1000	3–20	1

The complex permittivity ($\varepsilon_r = \varepsilon'$-$j\varepsilon''$) and permeability ($\mu_r = \mu'$-$j\mu''$) were measured from S-parameters by using the simulation program of the Reflection/Transmission Nicolson-Ross model.

6.2.2 Production of Polymer Composites with Hybrid Filler

Three types of nanoparticles were used as fillers in the prepared composites: multiwall carbon nanotubes (MWCNTs), barium hexaferrite ($BaFe_{12}O_{19}$) and magnetite (Fe_3O_4). Figure 6.2a, b, and c presents the electron microscopy of MWCNTs, $BaFe_{12}O_{19}$, Fe_3O_4 fillers respectively.

MWCNTs were chosen as conductive filler due to their low specific density, high thermal and corrosion resistances and excellent electrical conductivity. The form of these particles is similar to highly elongated cylinders. The used MWCNTs were purchased in CheapTubesIns, USA. Their dimensions are the following: diameter d is of 10–30 nm, length l is of 10 μm. As it is seen in Fig. 6.2a, initially MWCNTs are tangled. That's why at MWCNTs-based NC preparation dispersion of the composite mixture should be performed very careful. Table 6.1 summarizes the data on the size and morphology of the filler particles.

The powders of barium hexaferrite and spherical magnetite particles were used as secondary fillers in nanotubes/epoxy NC. The morphology of the mentioned magnetic fillers is essentially different in shape and size. As it is seen in Fig. 6.1b, barium hexaferrite powder consists of well-shaped plate-like hexagonal particles of

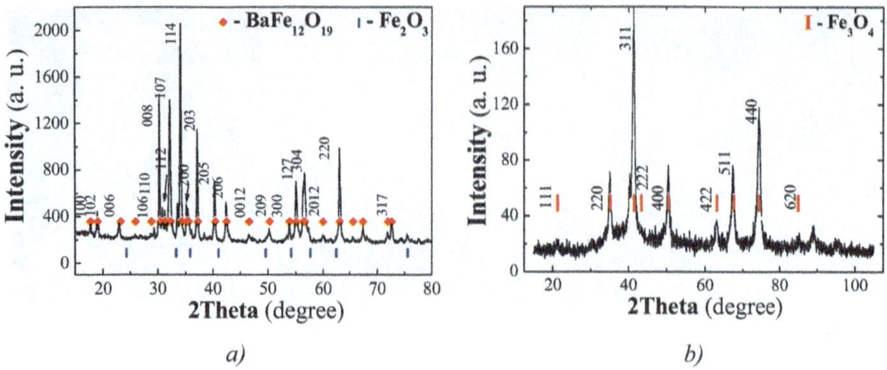

Fig. 6.3 XRD patterns for (**a**) $BaFe_{12}O_{19}$ (Cu K_α radiation, $\lambda = 1.541$ Å), (**b**) Fe_3O_4 (Co K_α filtered radiation, $\lambda = 1.7902$ Å)

20–50 nm thicknesses and 150–450 nm diameters. Particles size distribution is quite wide but hexagonal shape persists. Compact self-organization of barium hexaferrite particles takes place, and individual $BaFe_{12}O_{19}$ particles are gathered in globules where they stick at each other with random orientation of facets.

Figure 6.2c shows Fe_3O_4 powder of spherical particles of 10–20 nm diameters with minor dispersion in size.

Figure 6.3 shows X-ray diffraction (XRD) analysis results for the used secondary fillers. XRD patterns for ferrite powders confirm the presence of barium hexaferrite phase which possesses hexagonal lattice P63/mmc with a = 5.9046 Å, b = 5.9046 Å, c = 23.229 Å (indicated with symbols ♦ on Fig. 6.3a). XRD pattern of Fe_3O_4 is presented in Fig. 6.3b where the sharp diffraction peak at $2\Theta = 41.36°$ can be attributed to the (311) reflection of Fe_3O_4 nanoparticles. The main characteristic XRD peaks of magnetite are located at $2\Theta = 35.04°$, 50.47°, 67.38°, 74.30° and they correspond to the (220), (400), (511), (440) reflections of Fe_3O_4, respectively.

The series of NCs have been prepared with the above-mentioned fillers. Namely, MWCNTs/epoxy and MWCNTs/Fe_3O_4/($BaFe_{12}O_{19}$)/epoxy NCs with 2 wt. % and 5 wt. % of MWCNTs were fabricated by mixing in solution with additional sonication. The fabrication routine is as follows. At first, appropriate amount of L285 epoxy resin (Lange & Ritter GmbH, Germany) was pre-dissolved in acetone. Further, MWCNTs were introduced into solution and sonication in BAKU ultrasonic bath was applied for 15 min with 40 kHz frequency and 50 W power. Fe_3O_4 ($BaFe_{12}O_{19}$) was added at the next stage for the NC with binary filler. That secondary filler amounts to 30% of weight in all composites. Mechanical mixing and sonication for 15 min was applied again. Finally, curing agent H285 was added (40% by weight of L285).

At curing process, the prepared NCs were subjected to continuous ultrasonic (800 kHz) effect to provide uniform random distribution of fillers without agglomeration. To complete the polymerization, 1 day after the NCs were prepared, they were exposed to the heating at temperature that gradually increased from 40 to 80 °C for 5 h.

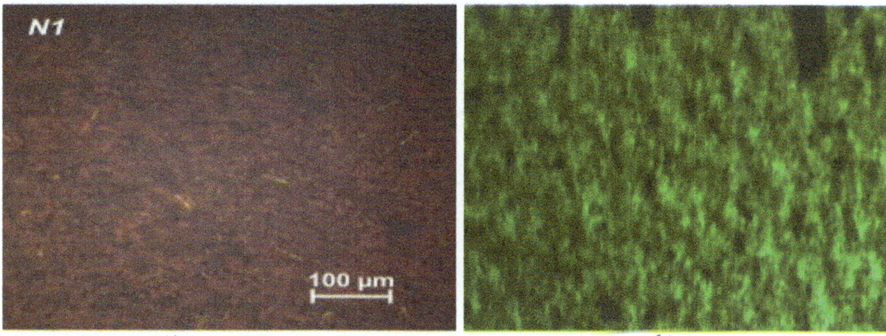

a) b)

Fig. 6.4 Optical images of (**a**) $BaFe_{12}O_{19}$/L285; (**b**) Fe_3O_4/L285 NC

The results of optical microscopic studies of $BaFe_{12}O_{19}$/epoxy, Fe_3O_4/epoxy NC surface are presented in Fig. 6.4.

As it is seen in Fig. 6.4 a, hexaferrite-filled NC is characterized by granular structure of $BaFe_{12}O_{19}$ globules distribution in epoxy matrix with 10–20 μm diameters of globules but there are many globules with a typical diameter of about 50 μm. These globules mostly have regular spherical form. Spatial orientation of $BaFe_{12}O_{19}$ globules is not observed. It should be noted that for composite samples with binary fillers the uniform distribution of the $BaFe_{12}O_{19}$ globules in the sample volume is slightly disturbed. Both $BaFe_{12}O_{19}$ phase and carbon phase form separate interpenetrating spatial networks. The nature of such formation in NCs with binary fillers can be linked to the influence of MWCNTs agglomerates on $BaFe_{12}O_{19}$ network formation. Specifically, van der Waals interaction forces between the carbon nanotubes prevent the $BaFe_{12}O_{19}$ phase distribution and contribute to its association.

A typical optical micrograph of Fe_3O_4/L285 NC is shown in Fig. 6.4b where Fe_3O_4 particles and their agglomerates are randomly oriented in the epoxy resin. Their distribution is rather uniform while agglomerates are present in large quantities.

Carbon nanoparticles are known as conductive structures while barium hexaferrite as well as ferric oxide are nonconducting. However, the addition of Fe_3O_4 or $BaFe_{12}O_{19}$ into the MWCNTs/epoxy NC leads to increase of electrical conductivity by 2 orders of magnitude for 2 wt. % MWCNTs/Fe_3O_4($BaFe_{12}O_{19}$)/L285 and by 3 orders of magnitude for 5 wt. % MWCNTs/Fe_3O_4($BaFe_{12}O_{19}$)/L285 NCs. These results are presented in Fig. 6.5.

However, as it was noted in [20], magnetite particles possess high interfacial resistivity and should not make a direct contribution to the increase of the composite electrical conductivity. Instead, they contribute to the formation of "excluded volume" in the composite, which results in the increase of local carbon filler concentration and promotes better dispersion of carbon filler at the manufacturing process. First of all, the composite electrical conductivity increase occurs due to

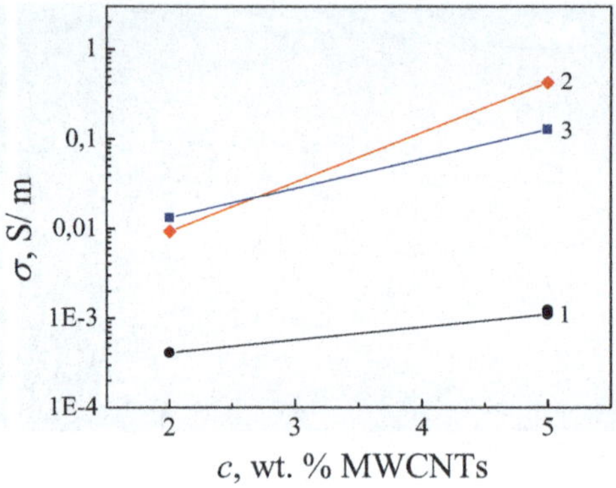

Fig. 6.5 Concentration dependences of conductivity of MWCNTs/$BaFe_{12}O_{19}$(Fe_3O_4)/L285 composites. Curves 1 – MWCNTs/L285; 2 – MWCNTs/$BaFe_{12}O_{19}$/L285; 3 – MWCNTs/Fe_3O_4/L285

the change in the distance between the conductive particles in the material when the second filler is introduced, i.e. due to the change of polymer layer thickness between the filler particles [21]. Authors have analyzed in [22, 23] in terms of the effective resistivity model the influence of a dielectric component addition on the percolation threshold and the electrical conductivity of NCs with carbon fillers of different types. It was shown that changes in those values depend on the shape and type of carbon particles used as conductive fillers. In the case of CNTs, the increase in electrical conductivity is mainly determined by the increase of the conductive chains number and the decrease of the contact resistance between the filler particles in the conductive chains.

6.3 Results and Discussion

6.3.1 Microwave Properties of Epoxy Composites with Carbon Nanotubes and Hybrid Fillers

The electromagnetic wave absorption are mainly determined by relative permittivity and permeability of the material in which the real parts of permittivity (ε') and permeability (μ') are associated with the electric and magnetic energy storage capabilities while the imaginary parts of permittivity (ε'') and permeability (μ'') stand for the electric and magnetic losses, respectively. In order to evaluate the influence of magnetodielectric particles types and content of MWCNTs on electrodynamic characteristics of hybrid polymer composites, complex permittivity and permeability were examined over a broad frequency range (1–67 GHz).

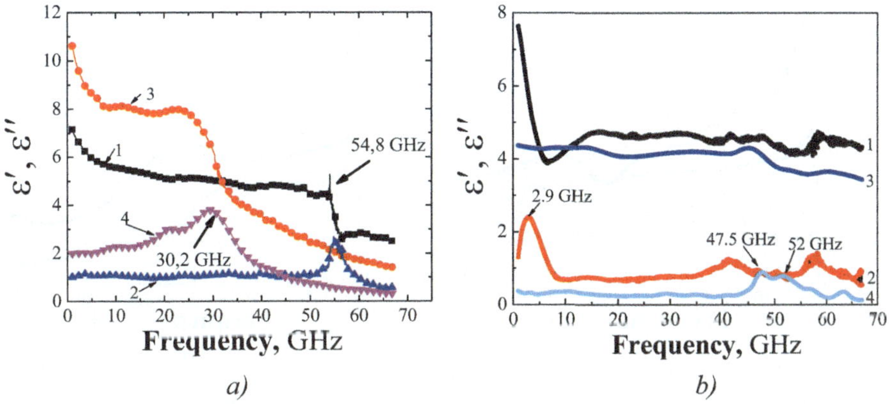

Fig. 6.6 Frequency dependences of real (ε') and imaginary (ε'') parts of complex permittivity for NCs: (**a**) MWCNT/L285; curves 1, 2 – ε' and ε'', curves 3, 4– ε' and ε'' for NCs with 2 wt. % MWCNTs/L285 and 5 wt. % MWCNTs/L285, respectively; (**b**) 30 wt. % $BaFe_{12}O_{19}(Fe_3O_4)$/L285; curves 1, 2 –ε' andε'', curves 3, 4– ε' and ε'' for 30 wt. % Fe_3O_4/L285 and 30 wt. % $BaFe_{12}O_{19}$/L285 CMs, respectively

6.3.1.1 Permittivity

The real and imaginary parts of permittivity in the range of 1–67 GHz for composites with individual fillers and for ternary composites are presented in Fig. 6.6.

As one can see in Fig. 6.6a, the significant rise of ε' is observed for MWCNTs/L285 NCs at the increase of MWCNTs content in the epoxy matrix. The composite with 5 wt. % MWCNTs shows higher ε' values as compared to the composite with 2 wt. % MWCNTs. With the higher nanotubes content the major peak of ε'' as well as the peak of dielectric loss factor are shifted towards lower frequencies. For instance, the major peaks of ε'' are observed at 54 GHz and 30.3 GHz for NCs with 2 wt. % and 5 wt. % of MWCNTs, respectively, suggesting a higher electrical conductivity for the samples with higher MWCNTs content.

For the comparison, the complex permittivities of NCs filled with solely $BaFe_{12}O_{19}$ or Fe_3O_4 particles are also presented in Fig. 6.6b.

It is notable that the real part of the complex permittivity of composites with an individual magnetodielectric filler is almost invariant for NCs with $BaFe_{12}O_{19}$ filler (up to 45 GHz) and for NCs with Fe_3O_4 filler at frequencies above 5 GHz, while the values of ε'' are near 1 and fluctuate as the frequency increases revealing several resonance peaks. ε' is the representation of the material dipolar polarizability at microwave frequencies, and ε'' curve resonance peaks demonstrate multi-relaxations also originated from the dipole polarization.

Figure 6.7 shows frequency dependences of complex permittivity for MWCNTs/$BaFe_{12}O_{19}$/epoxy NC and for MWCNTs/Fe_3O_4/epoxy NC with different MWCNTs content.

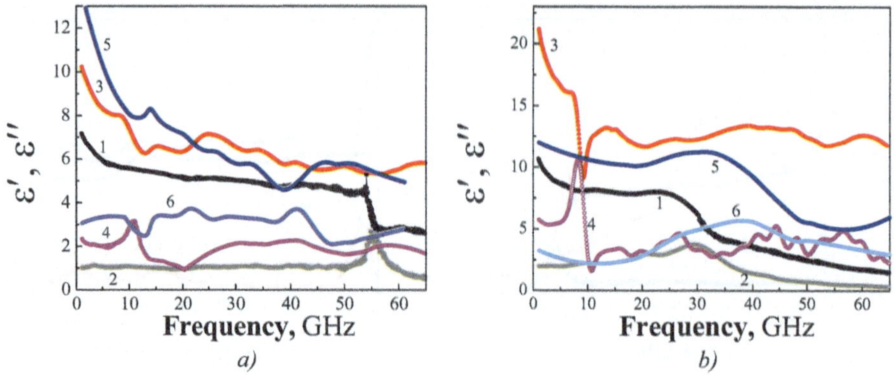

Fig. 6.7 Frequency dependences of real (ε') and imaginary (ε'') parts of complex permittivity of ternary composites MWCNTs/BaFeO(Fe_3O_4)/L285 with different MWCNTs content: (**a**) 2 wt. % MWCNTs/30 wt. % $BaFe_{12}O_{19}$(Fe_3O_4)/L285; curves 1, 3, 5 – ε' and curves 2, 4, 6 – ε'' for composites 2 wt. % MWCNTs/L285, 2 wt. % MWCNTs/30 wt. % Fe_3O_4/L285, 2 wt. % -MWCNTs/$BaFe_{12}O_{19}$/L285, respectively; (**b**) 5 wt. % MWCNTs/30 wt. % $BaFe_{12}O_{19}$–(Fe_3O_4)/L285; curves 1, 3, 5 – ε' and curves 2, 4, 6 – ε'' for composites 5 wt. % MWCNTs/L285, 5 wt. % MWCNTs/30 wt. % Fe_3O_4/L285, 5 wt. % MWCNTs/$BaFe_{12}O_{19}$/L285, respectively

Table 6.2 Influence of additional magnetic filler on permittivity of MWCNTs/epoxy NC

	$\frac{\varepsilon'_{CNT/Fe_3O_4}}{\varepsilon'_{CNT}}$		$\frac{\varepsilon'_{CNT/BaFe_{12}O_{19}}}{\varepsilon'_{CNT}}$		$\frac{\varepsilon''_{CNT/Fe_3O_4}}{\varepsilon''_{CNT}}$		$\frac{\varepsilon''_{CNT/BaFe_{12}O_{19}}}{\varepsilon''_{CNT}}$	
Frequency, GHz	20 GHz	60 GHz	20 GHz	60 GHz	20 GHz	60 GHz	20 GHz	60 GHz
2% MWCNTs	1,2	1,4	1,4	1,2	1,0	2,2	3,7	3,1
5% MWCNTs	1,5	7,0	1,3	3,0	1,1	8,5	1,0	7,0

The influence of additional magnetic filler on permittivity of MWCNTs/epoxy NC is analyzed in Table 6.2.

Substantial increase of both permittivity parts has been observed in ternary composites in comparison with binary composites (see Table 6.2). The distinct maximum have been observed for ternary composites with Fe_3O_4 at 2,6 GHz. Broad relaxation peak is appeared in the range of 40–45 GHz for 2 wt. % MWCNTs/$BaFe_{12}O_{19}$/epoxy composites which became more pronounced with the increase of MWCNTs content in the composite. The relative changes of ε' for composites with 2 wt. % of MWCNTs almost does not depend on the type of additional magnetic filler and frequency in whole investigated range. The relative changes of ε' for composites with 2 wt. % of MWCNTs are weakly dependent on the type of additional magnetic filler and frequency. The values of relative changes of real permittivity are approximately 1.2–1.4 while ε'' demonstrates higher changes which reflects substantial conductivity increase after addition of magnetic nanoparticles. On the contrary, significant frequency dependences of both permittivity parts have been observed for composites with higher MWCNTs content as well as the increase of ε' and ε'' in ternary composites in comparison with binary composites.

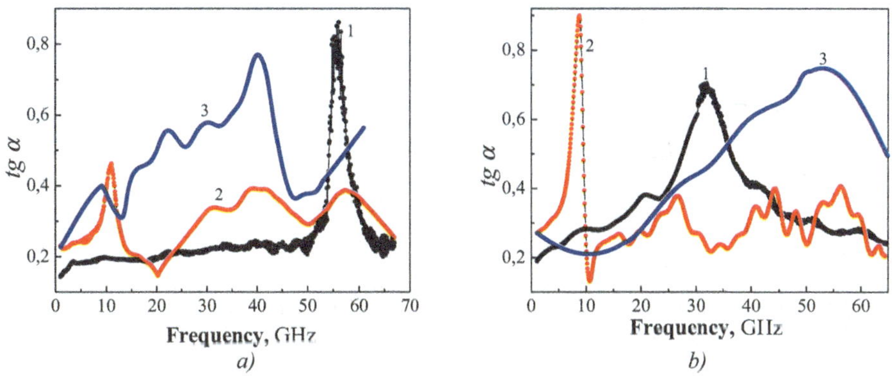

Fig. 6.8 Frequency dependences of*tg α* for ternary composites with different MWCNTs content: (**a**) 2 wt. %MWCNTs; (**b**) 5 wt. %MWCNTs; curves 1 – MWCNTs/L285, 2 – MWCNTs/Fe_3O_4/L285, 3 – MWCNTs/$BaFe_{12}O_{19}$/L285 respectively

The frequency-dependent dielectric loss tangents (tg $\alpha = \varepsilon''/\varepsilon'$) of the investigated NC are shown in Fig. 6.8. One can see that *tg α* in binary composites is about 0.2 almost in the whole frequency range and raises up to 0.7–0.9 in the relaxation peaks band.

For ternary composite with $BaFe_{12}O_{19}$ and Fe_3O_4 an essential enhancement of *tg α* was found the value of which increases with frequency.

Several relaxation peaks of dielectric loss tangent are manifested in Fig. 6.8 for both ternary and binary NCs. As it is known, dielectric properties of mixtures can be described in terms of the Maxwell–Wagner theory of interfacial polarization (interfacial polarization developed at conductor-insulator interface) where the accumulated electric charges increase with the introduction of conducting fillers.

The interfacial polarization and the associated relaxation always appear in a heterogeneous system, and the accumulation and uneven distribution of space charges at the interfaces will produce a macroscopic electric moment that can effectively absorb the incident EM energy. The relaxation process can be described by the Cole–Cole semicircle, which can be deduced based on the Debye dipolar relaxation. The relationship between ε' and ε'' is described as follows [24, 25]:

$$\left(\varepsilon' - \varepsilon_\infty\right)^2 + \left(\varepsilon''\right)^2 = \left(\varepsilon_s - \varepsilon_\infty\right)^2 \tag{6.1}$$

where ε_s is the static permittivity, and ε_∞ is the relative dielectric permittivity at the high-frequency limit. If there is a polarization relaxation process, the plot of ε'' versus ε' will be a single semicircle, which is also defined as the Cole-Cole semicircle.

Figure 6.9 shows typical $\varepsilon''(\varepsilon')$ curves for MWCNTs/L285 NCs (Fig. 6.9a), MWCNTs/Fe_3O_4/L285 NCs (Fig. 6.9b), and MWCNTs/$BaFe_{12}O_{19}$/L285 NCs (Fig. 6.9c).

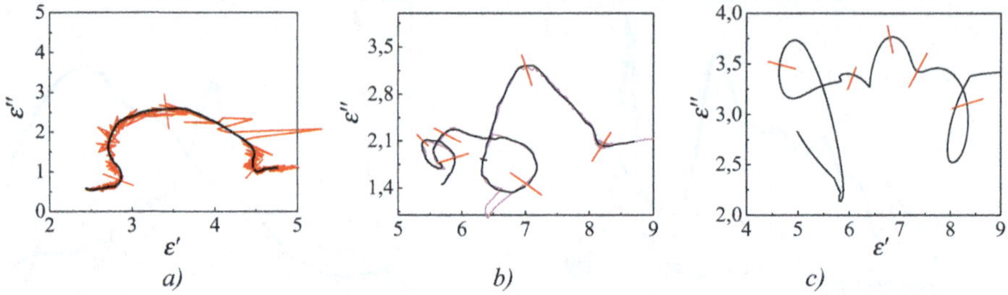

Fig. 6.9 Typicalε''-ε' curves of: (**a**) MWCNTs/L285 NC; (**b**) MWCNTs/Fe_3O_4/L285 NC; (**c**) MWCNTs/$BaFe_{12}O_{19}$/L285 NC

As it was observed, three semicircles for MWCNTs/L285 composites can be attributed to the Debye relaxation processes of L285 and nanotubes as well as to the interfacial polarization between nanotubes and epoxy. Obviously, five semicircles for MWCNTs/Fe_3O_4/L285 and MWCNTs/$BaFe_{12}O_{19}$/L285 composites can be attributed to the Debye relaxation processes of epoxy, nanotubes, and Fe_3O_4 or $BaFe_{12}O_{19}$ as well as to the interfacial polarization between epoxy, nanotubes, and magnetic filler. As one can see in Fig. 6.9a, the major peak of MWCNTs/L285 ε' is associated with interfacial polarization between nanotubes and epoxy as well as with the higher nanotubes content. This major peaks of ε'' and the dielectric loss tangent are shifted towards lower frequencies.

The major relaxation peak is located at 3.7 GHz for Fe_3O_4/L285 NCs and can be associated with interfacial polarization between Fe_3O_4 and epoxy. That peaks' position does not notably change after MWCNTs addition into a composite. The relaxation peaks are wide and feebly marked for $BaFe_{12}O_{19}$/L285 NCs that could be connected with the larger sizes of $BaFe_{12}O_{19}$ particles and their wider size distribution.

6.3.1.2 Permeability

The results of investigation of magnetic permeability of developed composites are presented in the Fig. 6.10.

Figure 6.10 shows the real part (μ') and the imaginary part (μ'') of the complex permeability of the developed composites. μ' of Fe_3O_4/L285 and Fe_3O_4-filled ternary composites is sharply-decreasing in the frequency range from 1 to 9 GHz while μ'' exhibits a peak located at ~2.01 GHz for Fe_3O_4/L285 NC. The latter could be caused by the natural ferromagnetic resonance (NFMR). It is known that NFMR of conventional Fe_3O_4 bulk materials is equal to 1.3 GHz [26].

Higher resonance frequency for Fe_3O_4/L285 NC can be explained by the natural resonance equation for ferromagnetic materials [27]:

$$f_r = \frac{\gamma}{2\pi}\frac{4\,|k_1|}{3\mu_0 M_s} \tag{6.2}$$

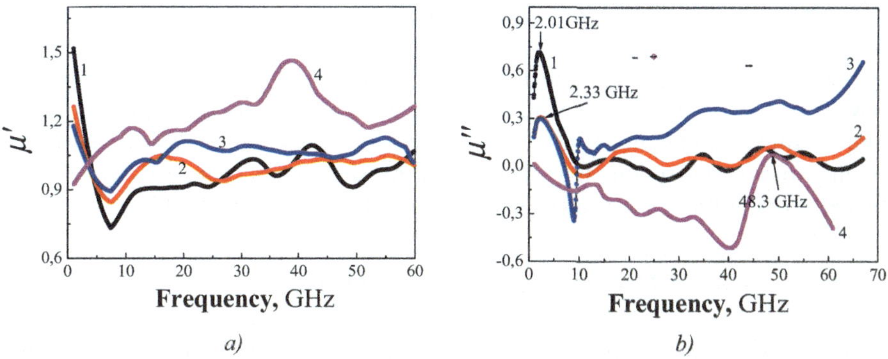

Fig. 6.10 (**a**) Real part (μ') and (**b**) imaginary part (μ'') of the complex permeability of developed NC: curves 1–30 wt. % Fe_3O_4/L285, 2–2 wt. % MWCNTs/30 wt. % Fe_3O_4/L285, 3–5 wt. % MWCNTs/30 wt. % Fe_3O_4/L285, 4–2 wt. % MWCNTs/30 wt. % $BaFe_{12}O_{19}$/L285

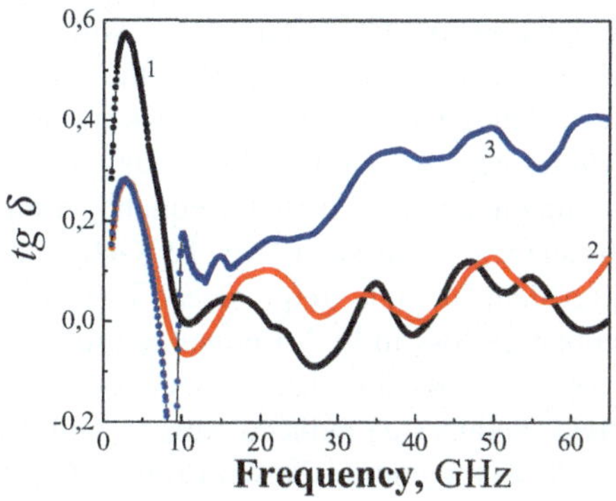

Fig. 6.11 Magnetic loss factor of NCs: curve 1–30 wt. % Fe_3O_4/L285, 2–2 wt. % MWCNTs/30 wt. % Fe_3O_4/L285, 3–5 wt. % MWCNTs/30 wt. % Fe_3O_4/L285

where γ is the gyromagnetic ratio, κ_1 is the anisotropy coefficient, μ_0 is permeability of free space, and M_s is the saturation magnetization .

As our previous $BaFe_{12}O_{19}$/epoxy polymer composite magnetic properties research [28] has shown, the encapsulation of magnetic powder in polymer core leads to changes of chemical bonding at the particles surface. That leads to decrease of the saturation magnetization of the magnetic particles and affects on the contributions of the surface anisotropy, the shape anisotropy, and the interface anisotropy to the net anisotropy. So, polymer coating of fine Fe_3O_4 particles and subsequent changes of their magnetic characteristics (particularly, saturation magnetization decrease) in a polymer composite produces the shift of f_r towards higher frequencies. MWCNTs addiction into the NC results in subsequent decreasing of M_s and shift of f_r towards higher frequencies which is reflected in initial increase of magnetic loss factor ($tg\ \delta = \mu''/\mu'$). $tg\ \delta$ frequency dependencies for the developed NCs are presented in Fig. 6.11.

μ' of $BaFe_{12}O_{19}$-based NCs is essentially higher than μ' of Fe_3O_4-based NCs, and μ'' demonstrates a peak with the maximum at 48.3 GHz. Such frequency is somewhat higher of the value that is known for the natural ferromagnetic resonance of bulk $BaFe_{12}O_{19}$ [29].

For magnetic EM absorption materials, the magnetic loss originate from the magnetic hysteresis, domain wall resonance, natural ferromagnetic resonance, exchange resonance, and eddy current effect [30]. The magnetic hysteresis loss is negligible in weak EM field. The domain wall resonance generally occurs in a lower frequency range in multi-domain materials. The eddy current loss is another important factor for microwave absorption. It is determined by the electric conductivity (σ) and thickness (d) of the samples and can be evaluated by the parameter [31]:

$$C = \mu''\left(\mu'\right)^{-2} f^{-1} - = 2/3\pi\mu_0\ \sigma d^2 \tag{6.3}$$

where μ_0 is the permeability in a vacuum, σ is the electric conductivity of the material. If C is a constant with the change of frequency, we can say that the magnetic loss results from the eddy current loss [31].

To analyze the effect of eddy-current on the magnetic loss, we plotted $\mu''(\mu')^{-2} f^{-1}$ via frequency for ternary composites with Fe_3O_4 and $BaFe_{12}O_{19}$. In case of MWCNTs/epoxy NC with Fe_3O_4, C decreases with frequency increase up to 12 GHz, implying that the eddy current has no significant effect on the electromagnetic microwave absorption in the frequency range below 12 GHz. Therefore, the natural ferromagnetic resonance may be the main contributor to magnetic loss in that frequency range. C becomes almost constant with the frequency increase above 12 GHz, i.e. eddy-current-induced magnetic loss is prevalent at high frequencies. The increase of non-magnetic MWCNTs content enhances loss in high frequency range where eddy-current-induced mechanism of magnetic loss dominates. A conductive material subjected to an alternating EM field produces an induced current, which dissipates EM energy within the material.

As to the composite with $BaFe_{12}O_{19}$ filler, as it is seen from Fig. 6.12, C is not constant and has serious fluctuations in whole studied field, indicating that the eddy

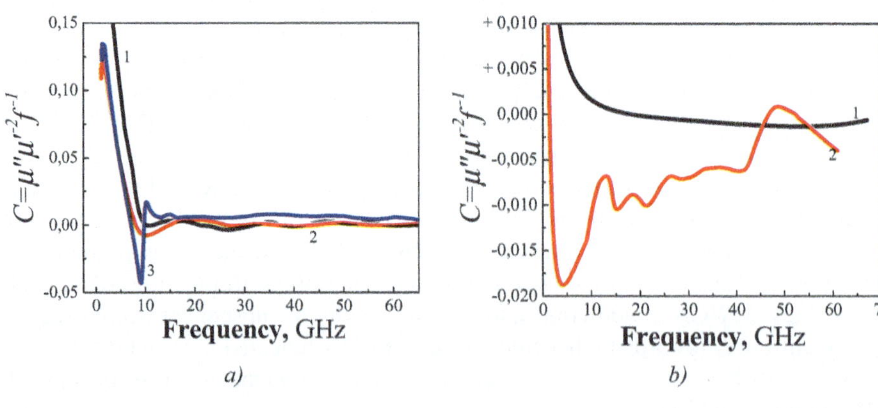

Fig. 6.12 Frequency dependences of C factor for ternary composites with Fe_3O_4 (**a**) (curve 1–30 wt. % Fe_3O_4/L285, 2–2 wt. % MWCNTs/30 wt. % Fe_3O_4/L285, 3–5 wt. % MWCNTs/30 wt. % Fe_3O_4/L285) and $BaFe_{12}O_{19}$ (**b**) (curve 1–30 wt. % $BaFe_{12}O_{19}$/L285, 2–2 wt. % MWCNTs/30 wt. % $BaFe_{12}O_{19}$/L285)

current effect has no significant effect on the microwave absorption observed and natural resonance loss determined the character of magnetic loss.

Interestingly, it is observed from Fig. 6.10 *b* that the $BaFe_{12}O_{19}$ NC show negative μ'' in the frequency range 9.5–18.0 GHz and the minimum value of μ'' is −0.17. In order to prove the reliability of the negative μ'', we measured the EM parameters of the samples which were produced in different production cycles and with different thicknesses and we found that μ'' was always negative in this frequency range, manifesting that μ'' was intrinsically negative rather than an artifact. The similar behavior of μ'' had been observed in [32] for nanocomposites containing 30 wt. % of iron nanorods. We have no reasonable explanation of observed results and more studies are necessary to understand the originality of the negative μ.

Thus, the investigated materials have shown feasible microwave properties for effective EM wave attenuation. The attenuation constant α is often used to take account of the integral dissipation [33]:

$$\alpha = \left(\sqrt{2}\pi f/c\right) \times \left(\left(\mu''\varepsilon''-\mu'\varepsilon'\right) + \left(\left(\mu''\varepsilon'' - \mu'\varepsilon'\right)^2 + \left(\varepsilon'\mu'' + \varepsilon''\mu'\right)^2\right)^{0,5}\right)^{0.5} \tag{6.4}$$

where f is the frequency of the EM wave and c is the velocity of light. Figure 6.13 shows frequency dependence of α of the investigated CMs.

In order to evaluate the efficiency of ternary composites as absorbing materials, their attenuation constants have been calculated using the experimental values of μ'',ε'', μ', ε'. The increase of α in MWCNT/L285 NCs has been observed with the increase of MWCNTs content. Besides it, the frequency range of maximum attenuation shifts to the lower frequencies. Naturally, the addition of magnetic nanoparticles into the composite leads to enhancement of attenuation. However, as Fig. 6.13 demonstrates, EMR attenuation is more intensive and realized in wider and higher frequency range in MWCNTs/$BaFe_{12}O_{19}$/L285 in comparison with MWCNTs/Fe_3O_4/L285 NCs at the same mass content of $BaFe_{12}O_{19}$ and

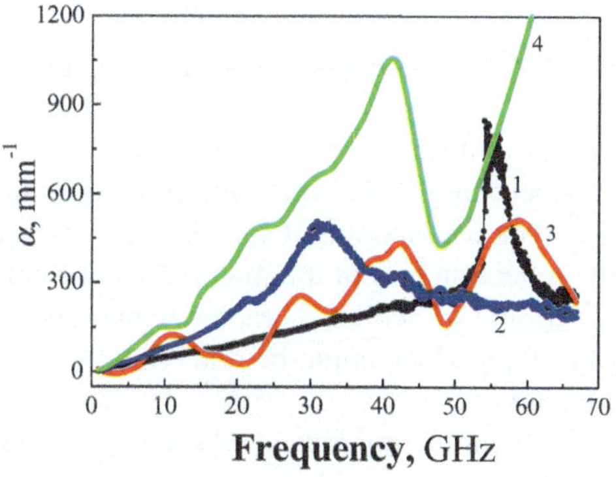

Fig. 6.13 Frequency dependences of attenuation constant α for ternary composites: curve 1–2 wt. % MWCNTs/L285; 2–5 wt. % MWCNTs/L285; 3–2 wt. % MWCNTs/30%Fe_3O_4/L285, 4–5 wt. % MWCNTs/30%Fe_3O_4/L285

Fe_3O_4 particles in the corresponding composites. It may be primarily connected with NFMR-based mechanism of magnetic loss in $BaFe_{12}O_{19}$ and consequently in $BaFe_{12}O_{19}$-based composites. Which is realized for $BaFe_{12}O_{19}$ at frequency of 47 GHz [29].

6.3.1.3 Microwave-Absorbing Properties

It is known that the performance of microwave absorption properties reflects frequency, effective bandwidth, and RL_{min} values, so it is meaningful to investigate materials with good microwave absorption properties (MAP) over a wide frequency range. To reveal the microwave absorption performance of developed composites, their RL values were calculated using the relative complex permeability and permittivity at a given layer thickness. In theory, the incident microwave power is divided into portions of the reflected and the absorbed ones at the interaction with metal-backed composite layer

$$\begin{gathered} P_{in} = P_R + P_A \\ RL = 10\log P_R/P_{in}, \qquad A = 10\ \log P_A/P_{in} \end{gathered} \tag{6.5}$$

where P_{in} is the power of the incident wave, P_R is the power of the reflected wave, P_A is the power of the absorbed wave, *RL* and *A* are the reflection and the absorption losses, respectively, measured in decibels (dB). Here the microwave absorbing efficiency can be evaluated from *RL*. The larger the absolute value of *RL* is, the stronger will be the wave-absorbing ability.

In order to investigate the possibility to widen the range of strong absorption and increase the absorbing peak value for the composites with low concentration of nanocarbon filler, we simulated *RL* (dB) by applying the transmission line theory. *RL* of EMR under normal wave incidence at the surface of a single layer material backed by a perfect conductor can be defined as [34]:

$$RL = 20\log\left|\frac{Z_{in} - Z_0}{Z_{in} + Z_0}\right|, \tag{6.6}$$

where $Z_0 = \sqrt{\mu_0/\varepsilon_0}$ – is the characteristic impedance of free space, $Z_{in} = \sqrt{\mu_0\mu/\varepsilon_0\varepsilon}\left(\tanh\left(i \cdot 2\pi f\sqrt{\mu_0\mu\varepsilon_0\varepsilon d}\right)\right)$ – is the input impedance at the interface of free space and material.

According to the foregoing Eqs. (6.5) and (6.6), the surface reflection loss of an absorber is a function of six characteristics parameters: ε', ε'', μ', μ'', f and d. Thus, the reflection loss of the material can be calculated from the measured values of ε', ε'', μ' and μ'' which characterize namely the property of material and proper value of f and d which could be related to, if it is possible to say, design and condition of functioning.

To examine microwave-absorbing properties of developed ternary hybrid/epoxy composites in the wide frequency range (1–67 GHz), we have calculated *RL* (dB) for

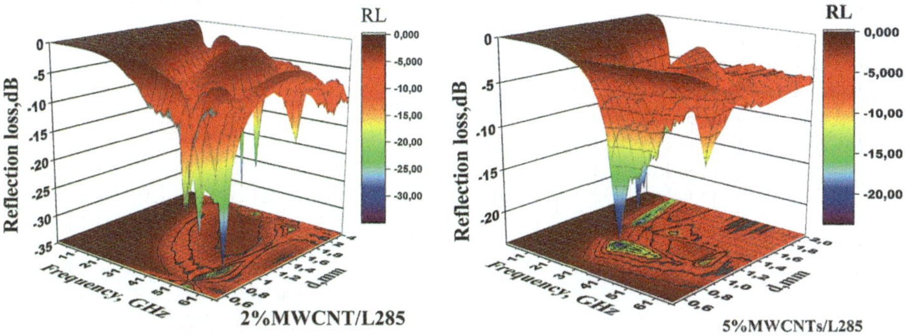

a) *b)*

Fig. 6.14 The reflection loss versus frequency and thickness for MWCNTs/L285 NC with 2 wt. % (**a**) and 5 wt. % (**b**) of MWCNTs

different thicknesses and different types of composites. It is important to analyze the influence of MWCNTs content and types of additional fillers on the *df* and RL_{min} values in order to obtain good MAP in wide effective frequency range, in particular, at higher frequencies with samples of minimum thickness.

The influence of MWCNTs content on MAP and above-mentioned absorber characteristics of MWCNTs/L285 composites is illustrated in Fig. 6.14. The results of *RL (f, d)* dependencies calculation are plotted for epoxy-nanotubes composites with 2 wt. % and 5 wt. % of MWCNTs.

The analysis of MA performance of composites with nanotubes have shown that increase of nanotubes content lead to worsening of MA characteristics. Namely, as one can see from the Figs. 6.14a, b, value of RL_{min} decreases from 31 dB for 2 wt. % MWCNTs/L285 NC to 23 dB for 5 wt. % MWCNTs/L285 NC. In the investigated frequency range, the effective reflection loss ($RL \leq -10$ dB) of 2 wt. % MWCNTs/L285 NC covers wider frequency range than that of 5 wt. % MWCNTs/L285 NC. The widest frequency range of effective reflection loss ($RL \leq -10$ dB) for 2 wt. % MWCNTs/L285 NC reach 6 GHz at $d = 0.8$ mm. The widest frequency range of effective reflection loss for 2 wt. % MWCNTs/L285 NC reaches 6 GHz at $d = 0.8$ mm. Furthermore, considerable value of effective reflection loss appears at higher frequencies of 50–57 GHz, and it was observed for samples with thickness $d < 1$ mm. It comes from the complex permittivity excessive increase with the nanotubes content increase while the complex permeability remains practically unchanged. Such modification leads to stronger impedance mismatch at the air-composite interface. Figure 6.15 shows *df* and f_{min} dependences on MWCNTs/L285 NCs thickness.

Besides the effect of nanotubes content, the additional magnetic filler content and type also have important effects on the MAP of hybrid composites. The character of MAP changes in composites depends on the additional magnetic filler: Fe_3O_4 or $BaFe_{12}O_{19}$. Figure 6.16 shows reflection loss versus frequency and thickness for the investigated ternary composites 2 wt. % MWCNTs/30 wt. % Fe_3O_4 (or 30 wt. % $BaFe_{12}O_{19}$).

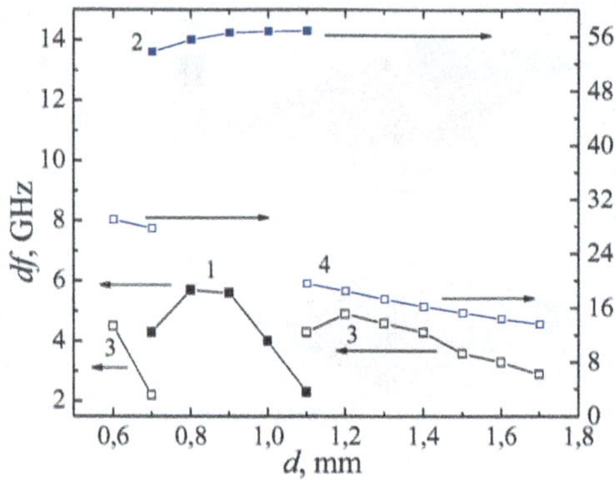

Fig. 6.15 The dependences of df and f_{min} versus thickness for MWCNTs/L285 NCs with 2 wt. % (curves 1, 2) and 5 wt. % (curves 3, 4) of MWCNTs

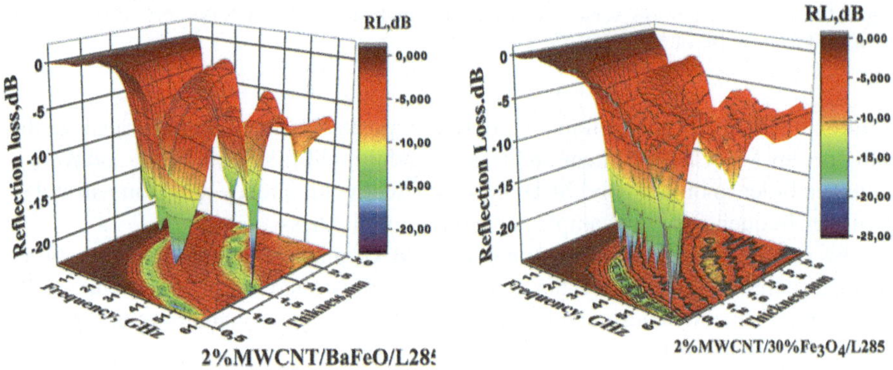

a) *b)*

Fig. 6.16 The reflection loss versus frequency and thickness for 2 wt. % MWCNTs/30 wt. % $BaFe_{12}O_{19}$/L285 (**a**) and 2 wt. % MWCNTs/30 wt. % Fe_3O_4/L285 (**b**) NCs

As one can see, the integration of nanotubes and magnetic inorganic additives such as Fe_3O_4 or $BaFe_{12}O_{19}$ creates obvious synergetic effects and complementary behavior as well as more loss mechanisms (e.g. various polarization relaxations), which greatly contribute to the MAP for both types of ternary composites. *df* reaches rather high value of 13–17 GHz for thin samples ($d = 0.6$–0,8 mm) of composites with 2 wt. % of nanotubes (see Fig. 6.17).

Such wide band of effective reflection loss ($RL > 10$ dB) is observed at $f > 20$ GHz. The value of reflection loss is not very high and does not exceed 20 dB for both composite types. $RL < 10$ dB was observed in the whole examined frequency range for rather thick (> 1.2 mm) samples of composites with Fe_3O_4 while RL of 6–11 dB was observed at high (30–60 GHz) frequencies for $BaFe_{12}O_{19}$-filled composite samples of 1–2 mm thickness.

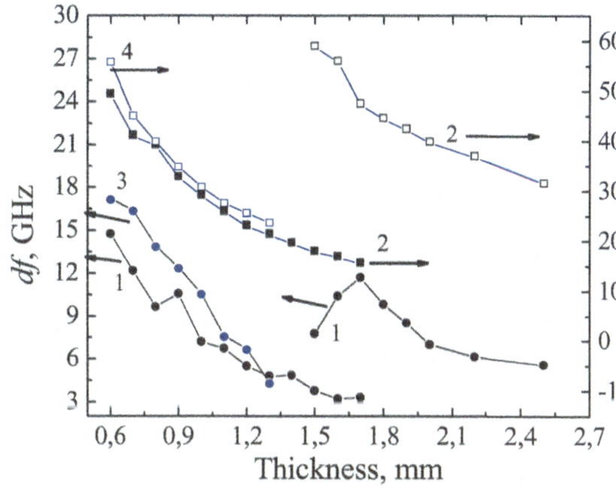

Fig. 6.17 The dependences of *df* and f_{min} versus thickness for composites 2 wt. % MWCNTs/30 wt. % $BaFe_{12}O_{19}$/L285 (curves 1, 2) and 2 wt. % MWCNTs/30 wt. %/Fe_3O_4/L285 (curves 3, 4)

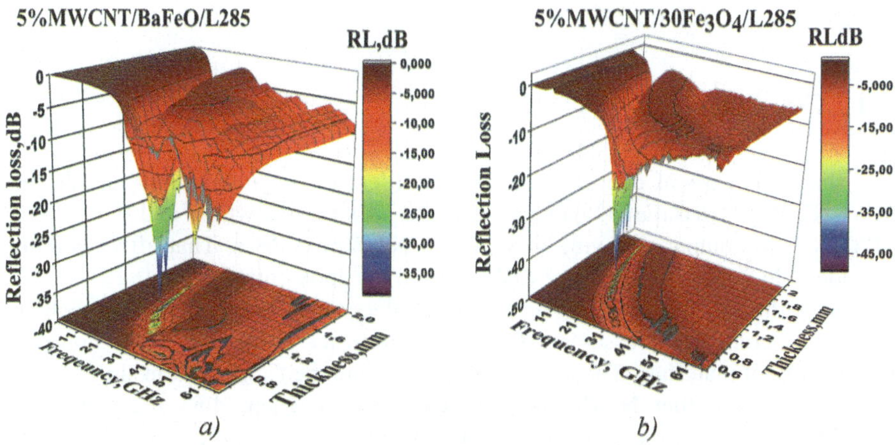

Fig. 6.18 The reflection loss versus frequency and thickness for: (**a**) 5 wt. % MWCNTs/30 wt. % $BaFe_{12}O_{19}$/L285 and (**b**) 5 wt. % MWCNTs/30 wt. % Fe_3O_4/L285 NCs

The reflection loss versus frequency and thickness for the investigated ternary composites 5 wt. % MWCNTs/30 wt. % Fe_3O_4 (or 30 wt. % $BaFe_{12}O_{19}$) NCs is presented at Fig. 6.18. Figure 6.19 illustrates the influence of MWCNTs content and additional filler type on composites microwave properties.

Increase of MWCNTs content in MWCNTs/L285 NCs as well as in the case of ternary compounds with additional Fe_3O_4 or $BaFe_{12}O_{19}$ leads to decrease of the effective reflection loss in the frequency range more than 20 GHz unless under increasing *d* up to 1,2–1,1 mm RL_{min} reach rather high value −49 dB and −39 dB for Fe_3O_4 and $BaFe_{12}O_{19}$ composite, consequently, but in this case *df* don't exceed 5 dB and f_{min} is founded in the region 20 GHz.

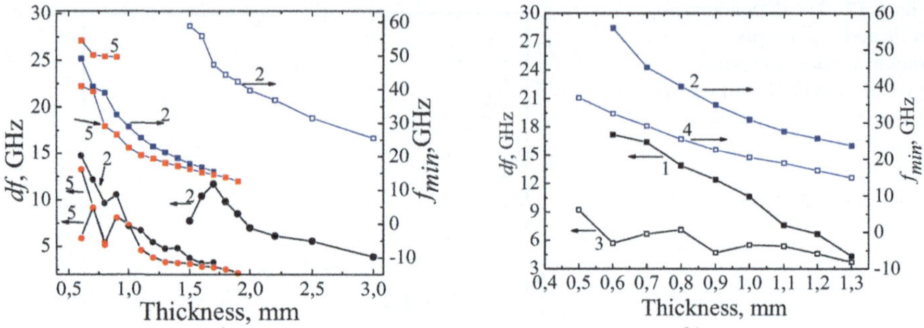

Fig. 6.19 The dependences of *df* and f_{min} versus thickness for composites: (**a**) 2 wt. % MWCNTs/30 wt. % BaFe12O19/L285 (curves 1, 2) and 5 wt. % MWCNTs/30 wt. % BaFe12O19/L285 (curves 3, 4); (**b**) 2 wt. % MWCNTs–/30 wt. % Fe_3O_4/L285 (curves 1, 2) and 5 wt. % MWCNTs/30 wt. % Fe_3O_4/L285 (curves 3, 4)

6.4 Conclusions

The sharp increase of conductivity of 3–5 orders of magnitude was revealed in MWCNTs/Fe_3O_4/($BaFe_{12}O_{19}$)/epoxy ternary NCs with 2 wt. % and 5 wt. % of MWCNTs in comparison with binary MWCNTs/epoxy NCs. That effect is more considerable for lower content of MWCNTs in the epoxy matrix.

The addition of 30 wt. % of Fe_3O_4 ($BaFe_{12}O_{19}$) nanoparticles as well as MWCNTs content increase in NCs have significant influence on the complex permittivity of materials. The ε_r values increase and the f_{min} position shifts.

It was shown that NFMR may be the main magnetic loss mechanism in MWCNTs/Fe_3O_4/epoxy composites at frequencies below 10–12 GHz while eddy current-induced magnetic loss prevails with the increase of microwave frequency. The natural ferromagnetic resonance determines the character of magnetic loss in $BaFe_{12}O_{19}$–filled NCs in the whole examined frequency range.

The analysis of microwave absorbing properties of the developed hybrid NCs has demonstrated good prospective of the NCs application as effective absorbing materials in the frequency range beyond 20 GHz.

High values of effective reflection loss (up to 17 dB) at frequencies above 20 GHz can be reached by the manipulating of MWCNTs content and types of additional magnetic fillers in hybrid NCs with *d* of 0.6–0.8 mm.

References

1. Golio M, Golio J (2007) RF and microwave passive and active technologies. CRC Press, London, p 736
2. Thomassin JM, Jerome C, Pardoen T, Bailly C, Huynen I, Detrembleur C (2013) Polymer/carbon based composites as electromagnetic interference (EMI) shielding materials. Mater Sci Eng R 74(7):211–232

3. Park DH, Lee YK, Park SS, Lee CS, Kim SH, Kim WN (2013) Effects of hybrid fillers on the electrical conductivity and EMI shielding efficiency of polypropylene/conductive filler composites. Macromol Res 21(8):905–910
4. Yoo TW, Lee YK, Lim SJ, Yoon HG, Kim WN (2014) Effects of hybrid fillers on the electromagnetic interference shielding effectiveness of polyamide 6/conductive filler composites. J Mater Sci 49(4):1701–1708
5. Chhowalla M, Shin HS, Eda G, Li L, Loh KP, Zhang H (2013) The chemistry of two-dimensional layered transition metal dichalcogenide nanosheets. Nat Chem 5(4):263–275
6. Hu C, Mou Z, Lu G, Chen N, Dong Z, Hu M, Qu L (2013) 3D graphene–Fe_3O_4 nanocomposites with high-performance microwave absorption. Phys Chem Chem Phys 15(31):13038–13043
7. Zhang H, Xie A, Wang C, Wang H, Shen Y, Tian X (2013) Novel rGO/α-Fe_2O_3 composite hydrogel: synthesis, characterization and high performance of electromagnetic wave absorption. J Mater Chem A 1(30):8547–8552
8. Zhang X, Wang G, Cao W, Wei Y, Liang J, Guo L, Cao M (2014) Enhanced microwave absorption property of reduced graphene oxide (RGO)-MnFe2O4 nanocomposites and polyvinylidene fluoride. ACS Appl Mater Interfaces 6(10):7471–7478
9. Wu F, Xia Y, Wang Y, Wang M (2014) Two-step reduction of self-assembed three-dimensional (3D) reduced graphene oxide (RGO)/zinc oxide (ZnO) nanocomposites for electromagnetic absorption. J Mater Chem A 2(47):20307–20315
10. Liu PB, Huang Y, Wang L (2013) Synthesis and excellent electromagnetic absorption properties of polypyrrole-reduced graphene oxide–Co_3O_4 nanocomposites. J Alloys Compd 573:151–156
11. Phan CH, Mariatti M, Koh YH (2016) Electromagnetic interference shielding performance of epoxy composites filled with multiwalled carbon nanotubes/manganese zinc ferrite hybrid fillers. J Magn Magn Mater 401:472–478
12. Li N, Huang GW, Li YQ, Xiao HM, Feng QP, Hu N, Fu SY (2017) Enhanced microwave absorption performance of coated carbon nanotubes by optimizing the Fe_3O_4 nanocoating structure. ACS Appl Mater Interfaces 9(3):2973–2983
13. Li J, Xie Y, Lu W, Chou T-W (2018) Flexible electromagnetic wave absorbing composite based on 3D rGO-CNT-Fe_3O_4 ternary films. Carbon 129:76–84
14. Kong B, Li ZW, Liu L, Huang R, Abshinova M, Yang ZH, Tang CB, Tan PK, Deng CR (2013) Recent progress in some materials and structures for specific electromagnetic applications. Int Mater Rev 58(4):203–259
15. Zahari MH, Guan BH, Meng CE, Mansor MFC, Chuan LK (2016) EMI shielding effectiveness of composites based on barium ferrite, PANI, and MWCNT. PIER M 52:79–87
16. Qing Y, Zhou W, Huang S, Huang Z, Luo F, Zhu D (2014) Evolution of double magnetic resonance behavior and electromagnetic properties of flake carbonyl iron and multi-walled carbon nanotubes filled epoxy-silicone. J Alloys Compd 583:471–475
17. Liu PB, Huang Y, Yan J, Yang YW, Zhao Y (2016) Construction of CuS nanoflakes vertically aligned on magnetically decorated graphene and their enhanced microwave absorption properties. ACS Appl Mater Interfaces 8:5536–5546
18. Biswas S, Kar GP, Bose S (2015) Tailor-made distribution of nanoparticles in blend structure toward outstanding electromagnetic interference shielding. ACS Appl Mater Interfaces 7(45):25448–25463
19. Gao Y, Gao X, Li J, Guo S (2017) Improved microwave absorbing property provided by the filler's alternating lamellar distribution of carbon nanotube/carbonyl iron/ poly(vinyl chloride) composites. Compos Sci Technol 158:175–185
20. Korskanov V, Dolgoshey V, Shantaliy T, Karpova I, Dragan K, Rukhaylo M (2017) Influence of an external magnetic field on the structure formation and transport properties of polymeric composites based on epoxy and Fe_3O_4. Ceram: Sci Life 3(36):28–33
21. Lazarenko A, Vovchenko L, Prylutskyy Y, Matzuy L, Ritter U, Scharff P (2009) Mechanism of thermal and electrical conductivity in polymer-nanocarbon composites. Mat-wissu Werkstofftech 40:268–272

22. Perets Y, Matzuy L, Vovchenko L, Prylutskyy Y, Scharff P, Ritter U (2014) The effect of boron nitride on electrical conductivity of nanocarbon-polymer composites. J Mater Sci 49(5):2098–2105
23. Matzui L, Vovchenko L, Perets Y, Lazarenko O (2013) Electrical conductivity of epoxy resin filled with graphite nanoplatelets and boron nitride. Mat-wissuWerkstofftech 44(2–3):254–258
24. Chen YH, Huang ZH, Lu MM, Cao WQ, Yuan J, Zhang DQ, Cao MS (2015) 3D Fe_3O_4 nanocrystals decorating carbon nanotubes to tune electromagnetic properties and EnhanceMicrowave absorption capacity. J Mater Chem A 3(24):12621–12625
25. Zhao B, Zhao WY, Shao G, Fan BB, Zhang R (2015) Morphology-control synthesis of a core-shell structured NiCu alloy with tunable electromagnetic-wave absorption capabilities. ACS Appl Mater Interfaces 7:12951–12960
26. Song NN, Yang HT, Liu HL, Ren X, Ding HF, Zhang XQ, Cheng ZH (2013) Exceeding natural resonance frequency limit of monodisperse Fe_3O_4 nanoparticles via superparamagnetic relaxation. Sci Rep 3:3161
27. Kittel C (1948) On the theory of ferromagnetic resonance absorption. Phys Rev 73:155–161
28. Yakovenko OS, Matzui LY, Vovchenko LL, Trukhanov AV, Kazakevich IS, Trukhanov SV, Prylutskyy YI, Ritter U (2017) Magnetic anisotropy of the graphite nanoplatelet–epoxy and MWCNT–epoxy composites with aligned barium ferrite filler. J Mater Sci 52(9):5345–5358
29. Walz F, Rivas J, Martínez D, Kronmüller H (1994) Influence of Ba content on the magnetic after-effect spectra in barium ferrites. Phys Status Solidi A 143:137–148
30. Wang GZ, Gao Z, Tang SW, Chen CQ, Duan FF, Zhao SC, Lin SW, Feng YH, Zhou L, Qin Y (2012) Microwave absorption properties of carbon nanocoils coated with highly controlled magnetic materials by atomic layer deposition. ACS Nano 6:11009–11017
31. Zhang H, Rao Y, Guo J, Qin G (2016) Multiple-phase carbon-coated $FeSn_2$/Sn nanocomposites for high-frequency microwave absorption. Carbon 96:972–979
32. Fan X, Guan J, Wang W, Tong G (2009) Morphology evolution, magnetic and microwave absorption properties of nano/submicrometre iron particles obtained at different reduced temperatures. J Phys D Appl Phys 42(7):075006
33. Liu XG, Jiang JJ, Geng DY, Han Z, Liu W, Zhang ZD (2009) Dual nonlinear dielectric resonance and strong natural resonance in Ni/ZnO nanocapsules. Appl Phys Lett 94(5):053119
34. Al-Ghamdi AA, Al-Hartomy OA, Al-Solamy FR, Dishovsky N, Mihaylov M, Malinova P, Atanasov N (2016) Natural rubber based composites comprising different types of carbon-silica hybrid fillers. Comparative study on their electric, dielectric and microwave properties, and possible applications. Mater Sci Appl 7(6):295–306

Chapter 7
Evolution of Structural and Magnetic Characteristics of Template Synthesized Nickel Nanotubes

Artem Kozlovskiy, Maxim Zdorovets, Daryn Borgekov, Milana Ibragimova, Ilya Korolkov, Alena Shumskaya, Maksim Kutuzau, and Egor Kaniukov

Abstract This work is devoted to the study of the relationship between synthesis conditions with structure and morphological peculiarities of nickel nanotubes. The influence of both synthesis regimes (temperature, potential correlation and deposition time) and template parameters (pore diameters) is estimated. Based on established regularities in the evolution of morphology and structural parameters, dynamics of the nickel nanotubes main magnetic characteristics is analyze.

Keywords Template synthesis · Ni nanotubes · Texture analysis · Electrical properties · Magnetic properties

7.1 Introduction

A great deal of international attention is focused on nanostructures (NSs) because they have unique properties and are promising objects for chemical and biological separation [1], catalysis [2] and as biomedical agent [3, 4], as well as sensing elements of magnetic-field detectors [5–9]. To date, a wide variety of methods to produce NSs have been developed, such as the spontaneous coalescence of nanoparticles [9], hydrothermal synthesis [10], electrodeposition [11, 12], an electron-beam lithography [13], chemical vapor deposition [14], pulsed laser deposition [15], and some other methods [16–19]. The method of electrochemical deposition is perfect

A. Kozlovskiy · M. Zdorovets · D. Borgekov
Gumilyov Eurasian National University, Astana, Kazakhstan

Institute of Nuclear Physics, Almaty, Kazakhstan

M. Ibragimova · I. Korolkov
Institute of Nuclear Physics, Almaty, Kazakhstan

A. Shumskaya (✉) · M. Kutuzau · E. Kaniukov
Scientific-Practical Materials Research Centre, NAS of Belarus, Minsk, Belarus
e-mail: Lunka7@mail.ru

A. Maffucci, S. A. Maksimenko (eds.), *Fundamental and Applied Nano-Electromagnetics II*, NATO Science for Peace and Security Series B: Physics and Biophysics, https://doi.org/10.1007/978-94-024-1687-9_7

for metallic nanostructures synthesis, because it allows to efficiently manage the physical and chemical properties of the nanostructures with high degree of process control [20–22] by varying electrolyte composition, deposition potential and time as well as template parameters. Choice of this method give a great opportunity to produce NSs with different dimensions, appropriate for any commercial tasks. Martin, who was a pioneer of template synthesis, showed the versatility of the approach to producing nanostructure arrays that involved the use of porous matrices based on alumina [23], polycarbonate or polyethylenterephtalate [24, 25].

Unique physicochemical, magnetic, and electric properties of ferromagnetic NSs based on the iron-triad elements, such as Fe, Co, and Ni [26], as well as on alloys of these elements, explain the great interest of researchers and manufacturers to these NSs. Note that the properties of NSs are largely determined by both their material and shape. For example, a high aspect ratio of elongated nanosized objects enables the easy magnetic axis to be identified in these objects and the use of hollow NSs, e.g., nanotubes (NTs), makes it possible to produce NSs with uniform commutation fields, which ensure the reproducibility of the results, due to the absence of a magnetic core in these NSs. A low bulk density of hollow NSs makes them floatable in liquids, including biological ones, and applicable in biotechnology, e.g., as a base for producing contrast agents. An expanded specific area of the surface of these NSs enables a larger number of functional links to be established and, therefore, a larger amount of effective weights to be supplied in the case of drug delivery.

Taking into account the potential advantages of magnetic NTs for practical applications, as well as the lack of information about the systematic study of their structural and magnetic characteristics, in the present work a comprehensive study of the mechanisms of the nickel NTs formation in the pores of ion-track membranes at different synthesis conditions is provided and the relationship between the morphology and the structure of magnetic NTs with their magnetic correlation parameters is considered.

7.2 Experimental Part

Nickel NTs were formed in pores of polyethylenterephtalate (PET) ion- track membranes with thickness of 12 μm and 100–500 nm diameters with pores density of 4×10^7 cm^{-2} [27]. Electrodeposition was carried out in the electrolyte of the composition $NiSO_4 \cdot 6H_2O$ (100 g/L), H_3BO_3 (45 g/L), and $C_6H_8O_6$ (1.5 g/L) under potentiostatic mode at cathode potensial difference of 1.25–2.00 V and temperatures 25, 35, and 50 °C. The cathode was a 10-nm-thick gold film deposited using vacuum magnetron sputtering. This thickness of the gold film enabled pores to remain opened, and the growth of the NTs began at sites of the contact of gold with the electrolyte. The electrodeposition process was controlled using chronoamperometry implemented in an Agilent 34410A multimeter instrument.

The structure and morphology of the Ni NTs were studied using scanning electron microscopy (SEM) implemented in a Hitachi TM3030 instrument equipped with a Bruker XFlash MIN SVE energy-dispersive spectra analyzer (EDS) at an accelerating voltage of 15 kV. Inner diameters of the NTs formed in the PET templates were measured using the manometric method of determining gas permeability, which was based on measuring the gas pressure in a closed chamber in the range of 8–20 kPa with a step of 4 kPa. X-ray diffraction structural analysis (XRD) was performed using a D8 ADVANCE diffractometer equipped with an X-ray tube with a copper anode. XRD patterns were recorded in an angular range of $2\theta = 30°–100°$ with a step of 0.02°.

Magnetic measurements were performed by using the methods of vibration magnetometry by the vibrational magnetometer (Cryogenic Ltd). The investigations were made in magnetic fields up to 2 T and in temperature range 300 K.

7.3 Influence of Deposition Parameters on Morphology and Structure of Nickel Nanotubes

The simplest way to control the morphology and structure of metallic NSs during electrodeposition is the use of porous templates. The most common templates are: anodic aluminum oxide [28, 29], porous silicon [30] or polymeric membranes [31]. Polymer membranes are of interest for creating metal NSs on a elastic basis for use in the field of flexible electronics. The template parameters, as well as the deposition conditions make it possible to create metallic NSs with various shapes and sizes [32] (Fig. 7.1).

Variation of the deposition parameters, as well as the parameters of the template, leads to a change in the deposition rate. By varying the deposition rate, the composition, morphology, and structural parameters of the resulted NSs can be controlled. The growth of the deposition rate can be caused by an increase in the temperature and the potential difference of deposition, as well as a decrease in the pores diameter of the template. In addition, the deposition rate will depend on the concentration of metal ions in the electrolyte, and the acidity of the electrolyte will allow stabilizing metal ions in the solution and, corresponded deposition rate.

Control of the deposition time makes it possible to synthesize NSs with different lengths and, consequently, an aspect ratio. Among the NSs the most interesting for practical applications are nanowires (NWs), nanotubes (NTs) with an aspect ratio of more than 10, nanorods and nanorings with an aspect ratio of less than 10. The length of electrodeposited NSs corresponds to the stages of growth and determines the structural and physical properties.

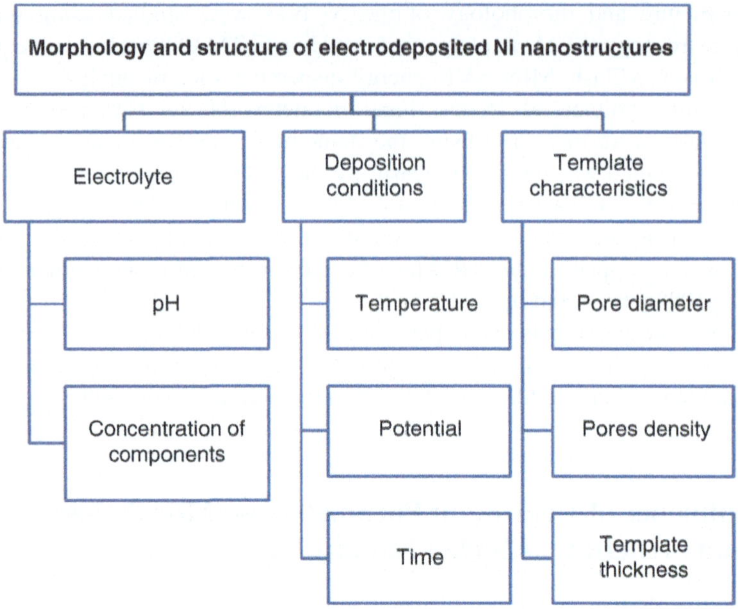

Fig. 7.1 Influencing factors on morphology and structure of nickel nanotubes by the template synthesis method

7.3.1 Peculiarities of the of Nickel Nanotubes Formation at Different Deposition Stages

A typical chronoamperogram of electrochemical deposition of Ni NTs in the pores of the PET template is presented in Fig. 7.2a. This type of I(t) dependence corresponds to four well-known stages of the growth process of NSs in the pores of the templates [33–34]: (I) nucleation (up to 50 s), (II) active growth of NSs (50–550 s), (III) formation of «caps» on the NSs upper part (550–700 s) and (IV) the deposition of the continuous metallic film on the template surface (more than 700 s, this section is not shown on the chronoamperogram). In order to study of the morphology evolution, the following deposition times were chosen: (1) – 80 s, (2) – 220 s, (3) – 360 s, (4) – 500 s, and (5) – 650 s. The sample obtained at time (1) corresponded to the deposition (stage I-II, nucleation and the beginning of active growth); samples (2)–(4) – formation of the main part of the NTs (stage II, active growth); sample (5) – corresponded to the growth of "caps" on NTs upper part (stage III). Schematic section of NSs formation at times (1)–(5) is shown in Fig. 7.2b.

SEM images of the samples are shown in Fig. 7.2c. The synthesized NSs have an outer diameter corresponding to the pore diameter of the template 380 $\pm$ 10 nm. The NSs obtained at time (5) have characteristic outgrowths in the upper part ("caps"). Analysis of SEM images allows establishing the presence of hollow internal channel in the NSs, i.e. the NTs. The internal diameters have been measured by gas permeability.

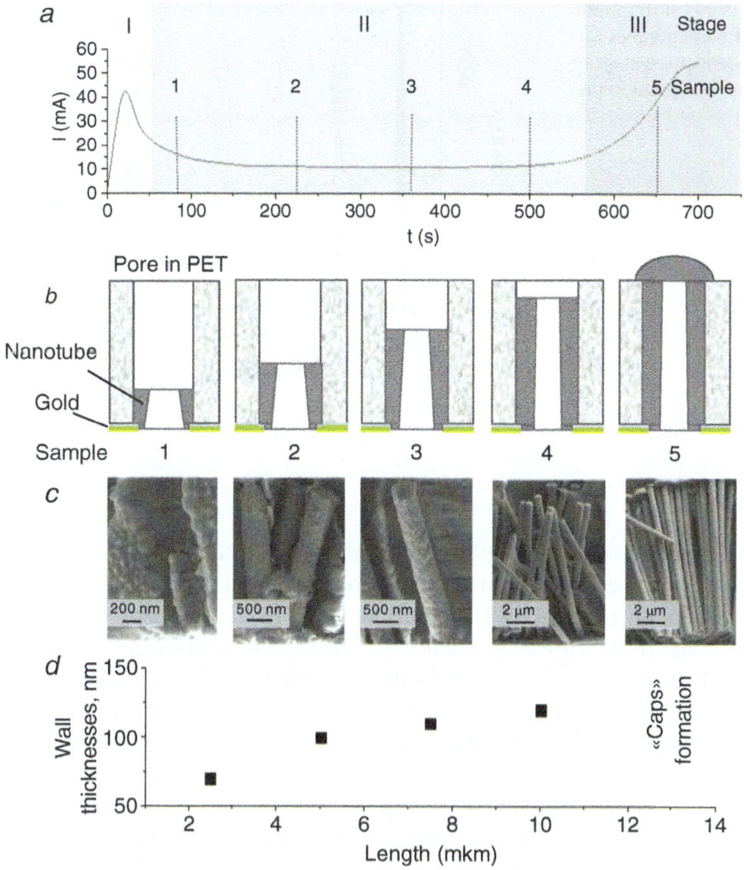

Fig. 7.2 Ni nanotubes growth process: I(t) dependence (**a**), schematic representation (**b**), typical SEM images (**c**), wall thicknesses dependence (**d**)

The increase of NTs length provides the inner diameter decrease, correspondingly, the wall thickness growths (Fig. 7.2d). The analysis of the NTs array SEM images of after template removing, the lengths of obtained NSs at different synthesis times have determined: (1) –2.8 ± 0.4 μm, (2) –5 ± 0.4 μm, (3) –7.3 ± 0.4 μm, (4) –10.5 ± 0.4 μm, (5) –12.5 ± 0.4 μm ("caps" are taking into account). The transformation in NTs morphology could be explained by discussing the deposition process.

The I stage corresponds to the onset of the filling of template pores. When the voltage is supplied, nuclei of the metal appear on the ring electrode in the bottom part of a pore, which replicate the shape of the electrode and set the type of the NSs being synthesized. Nuclei appear simultaneously in active zones on the surfaces of the pore walls, which contain defects, dangling bonds, etc.; these active zones result from chemical or physical activation, e.g., in the course of pore formation in

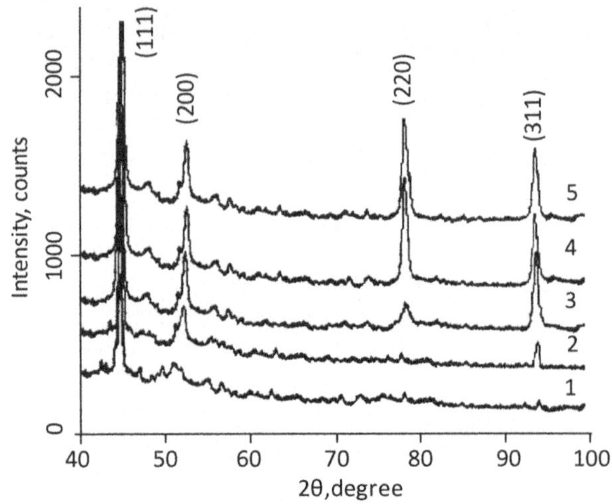

Fig. 7.3 X-ray diffraction spectra of Ni nanotubes at different deposition stages. Numbers correspond to the Fig. 7.2c

ion track-etched membranes and UV sensibilization. The intensive formation of the nuclei reduces the concentration of metal ions inside the pore and, therefore, leads to a substantial decrease in the plate current (see Fig. 7.2a).

At the stage II, the current gradually decreases or remains nearly constant, which corresponds to the growth of the NTs directly inside the pores until the NTs reach the surface of the membrane. Since the diffusion of ions into the pore occurs fairly slowly, the NTs grow at the expense of the successive overlaying of two-dimensional nuclei onto the surface of the ring.

For metal ions involved in NTs growth to reach the cathode, they should cross the depleted zone, i.e., pass from the high-concentration zone to the low-concentration zone through the narrow pore channel. Since the velocity of the movement of the ions toward the active zone is determined by the conditions of electroplating, an increase in voltage and the temperature promotes fast growth of NSs and formation of hollow channel. The reason is the gas intensification and its accumulation inside the NTs and hampers the ingress of the metal ions therein, which facilitates deposition near the pore walls.

Determination of the phase composition and crystalline structure was provided by X-ray diffraction (Fig. 7.3). The crystal structure of NTs was assessed on the basis of the X-ray diffraction spectra and it was shown all samples have F.C.C. crystalline structure.

To estimate the structure changes parameters, such as texture coefficients (TC_{hkl}), crystal lattice parameter (a), dislocation density (δ), average crystal size – esumed as coherent scattering region and degree of crystallinity were calculated. Texture coefficients TC_{hkl} are summarized in Table 7.1.

The number of reflections (n) corresponds to maximum value of texture coefficients. Analysis of TC_{hkl} shows the main crystalline direction [111], which is oriented priority along the axis of NTs. Studying diffractograms at the first minutes of synthesis (sample (1)) there is low-intensity peak with Miller indices (200)

Table 7.1 Texture coefficients of Ni nanotubes at different deposition stages

		TC_{hkl}				
2θ°	(hkl)	1	2	3	4	5
44.505	(111)	1.4131	1.4721	1.4874	1.5140	1.5211
51.844	(200)	0.3341	0.5231	0.6427	0.9983	1.1562
76.366	(220)	–	–	0.3532	0.4351	0.7531
92.939	(311)	0.1214	0.2313	0.5224	0.5854	0.6531

Table 7.2 Main structural parameters of Ni nanotubes at different deposition stages

Sample	Lattice parameter, Å	Coherent scattering region, nm	Crystallinity, %	δ, m2·1015
1	3.5243 ± 0.0011	9.3 ± 0.7	63	11.57
2	3.5210 ± 0.0020	12.2 ± 0.8	74	9.61
3	3.5192 ± 0.0008	14.5 ± 1.2	88	6.71
4	3.5141 ± 0.0023	17.1 ± 1.0	91	4.72
5	3.5131 ± 0.0011	21.9 ± 1.2	93	3.41

and (311). The broadening of these peaks for the (2) indicates the presence of microdistortions. With the synthesis time increasing in diffraction patterns the growth of intensity and narrowing of the (311) and (200) peaks can be seen, as well as the appearance of additional peaks with Miller indices (220), which indicates the presence of a crystalline texture. For (5), the [200] direction is singled out, which indicates a change in the crystal structure in the transition from tubular morphology to the "caps" formation. Dynamic of structural parameters of Ni NTs in growth process are presented in Table 7.2.

With increasing of synthesis time, a growth in the average crystallite size is observed, which is associated with nucleation processes and the formation of individual crystallites, which is subsequently form the wall of the tube. The increase in the size of the crystallites leads to the texturing of the tube during the synthesis.

7.3.2 *Influence of Deposition Potential on Morphology and Structure of Nickel Nanotubes*

Results of Ni NTs synthesis under different deposition potentials are shown on Fig. 7.4. The rate of deposition usually define the quality of the structure and in our case have influence on the wall thickness. At 1.25 V potential difference, structures are appeared with very small channel inside, while an increase in the potential difference leads to the hollow NSs formation with a tube shape. With increasing of deposition potential the chanel diameter is increasing and at 2.0 V have maximum value. It should be mentioned that at highest applied potential porous inclusions in NTs wall are observed, which may be due to the influence of the impurity on the formation of the NTs structure.

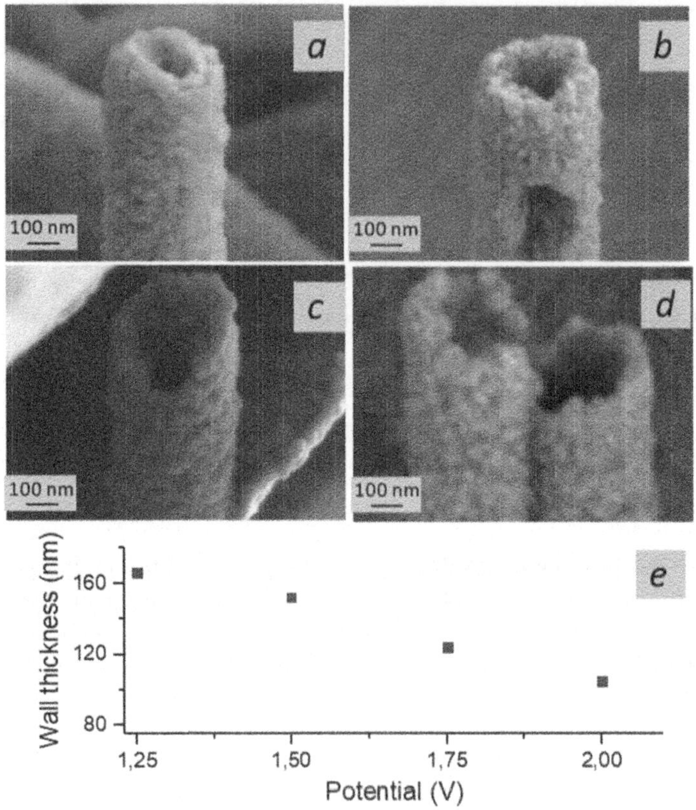

Fig. 7.4 SEM images of Ni nanotubes synthesized at different deposition potentials: (**a**) – 1.25 V; (**b**) – 1.5 V; (**c**) – 1.75 V; (**d**) – 2.0 V. (**e**) – dependence of wall thickness on deposition potential

X-ray diffraction analysis method was used to understand the influence of deposition potential on changes in structure and phase composition (Fig. 7.5). The estimated crystallinity and main structural parameters of Ni NTs synthesized at different deposition potentials are presented in Tables 7.3 and 7.4, correspondingly.

The obtained values of TC_{hkl} confirm the previously stated assumption of the polycrystalline structure of investigated samples. In this case, for the samples obtained at 1.25 V, the dominant textural plane is the plane (220). An increase in the applied potential difference leads to a rearrangement of the texture directions and the dominant plane becomes (111). In this case, the change in FWHM of the main diffraction lines indicates a change in the degree of crystallinity of synthesized samples obtained at various synthesis conditions (the results of the change in the degree of crystallinity are shown in Table 7.4).

The increase in crystal lattice parameter and the average size of crystallites can be explained by a change in the concentration of defects and microdistortions in the structure that are formed during the synthesis. An estimate of the diffraction maxima position indicates that when the synthesis conditions change. However, at 2.0 V

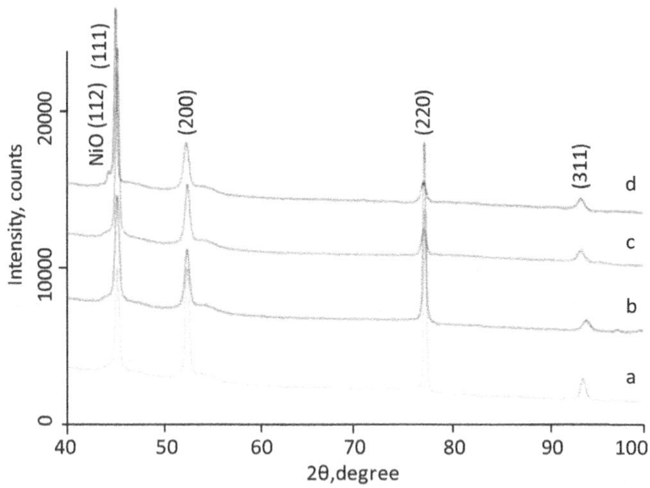

Fig. 7.5 X-ray diffraction patterns of Ni nanotubes synthesized at different deposition potentials: (**a**) – 1.25 V; (**b**) – 1.5 V; (**c**) – 1.75 V; (**d**) – 2.0 V

Table 7.3 Texture coefficients of Ni nanotubes synthesized at different deposition potentials

	TC_{hkl}			
Deposition potential, V	(111)	(200)	(220)	(311)
1.25	0.9879	1.2053	2.0352	0.5162
1.5	1.3525	0.7743	1.5352	0.4532
1.75	1.8979	0.6535	1.0433	0.3254
2.0	2.3892	0.6162	0.6463	0.2145

Table 7.4 Main structural parameters of Ni nanotubes synthesized at different deposition potentials

Deposition potential, V	Lattice parameters, Å	Coherent scattering region, nm	Crystallinity, %	δ, m2·1015
1.25	3.5141 ± 0.0021	15.9 ± 0.9	82	0.39
1.5	3.5197 ± 0.0018	19.7 ± 1.1	85	0.26
1.75	3.5237 ± 0.0027	26.0 ± 1.0	88	0.15
2.0	3.5372 ± 0.0019	41.4 ± 1.4	75	0.06

potential difference, there is an additional low intensity peak characteristic of nickel oxide with Miller indices (112). The appearance of the NiO peak is provided by the presence of impurities in the structure, which arise during the synthesis process at large values of potential. Oxygen enters to the crystal lattice during the electrolysis process and deforms it. A sharp increase in the size of crystallites with 2.0 V applied potential is due to the presence of oxide compounds impurities in the structure. Also, an increase of potential leads to a change in the texture of synthesized samples: for example, to a decrease in the intensity of diffraction peaks (200) and (220), and also to an increase in the intensity of the peak (111).

The degree of crystallinity increases with the rise of the potential from 1.25 to 1.75 V leads to the improvement of the crystal structure, while according to the energy dispersive analysis, no oxygen is observed in the structure. A sharp decrease in the degree of crystallinity is observed at 2.0 V. It happens because of oxide impurities presence in the structure, which leads to nanotubes partial amorphization, which confirms the results of SEM images (Fig. 7.4d).

One of the essential factors performs the crystal structure of NTs is the concentration of dislocations and stresses in the structure. Their change directly affects the physical characteristics. An increase in the degree of crystallinity as a result of changes in deposition conditions corresponds to a decrease in microstresses and deformation in the structure. However, the appearance of oxide impurities leads to a sharp increase in microstresses and defects in the crystal structure.

7.3.3 *Influence of Deposition Temperature on Morphology and Structure of Nickel Nanotubes*

Electrodeposition was carried out at various temperatures for deposition potentials 1.5 and 2 V. According to the SEM data, the overall dimensions of the NTs were estimated as 12 μm long and outer diameter of 400 nm. An analysis of inner diameters of the formed NTs inner diameter using the gas permeability technique was proved and the wall thickness was estimated. It was shown that an increase in the potential or the temperature leads to a decrease in the thickness of the walls from 150 to 60 nm (Fig. 7.6).

The XRD study of the phase composition of the nickel NTs (Fig. 7.7) has shown that in compression with results shown in the Sect. 3.1 the presence of NiO phase was not estimated. The XRD patterns recorded at diffraction angles $2\theta < 35°$ and $2\theta = 53°$ contain the peaks typical of PET films.

An analysis of the XRD data obtained under different conditions of synthesis has shown that nickel NTs produced at high potentials are characterized by an increase in the intensity of the (111) peak and a decrease in the intensity of the (200) peak. The observed redistribution of the intensities of the XRD reflections results from the appearance of the preferred orientation of the normals with regard to the (111) planes parallel to the normal to the plane of the specimen, i.e., in our case along

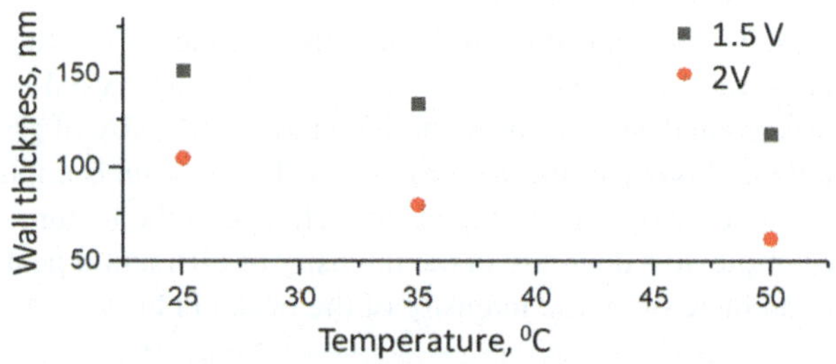

Fig. 7.6 Dependences of Ni nanotubes wall thickness on the deposition temperature at different potentials

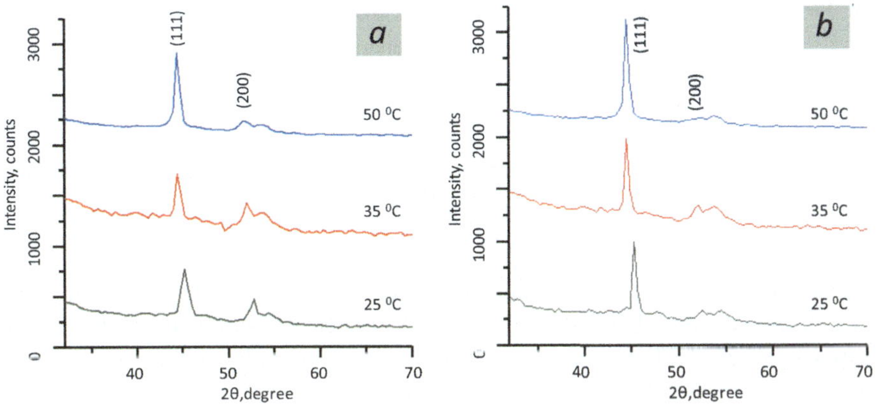

Fig. 7.7 XRD spectra of Ni nanotubes synthesized at different temperatures at potentials (**a**) 1.5 V and (**b**) 2 V

Table 7.5 Texture coefficients of Ni nanotubes synthesized at different temperatures

			TC_{hkl}	
2θ°	(hkl)	*T*, °C	1.5 V	2.0 V
44.559	(111)	25 °C	1.4762	1.7831
51.930	(200)		0.8224	0.5421
44.559	(111)	35 °C	1.5413	1.8741
51.930	(200)		0.7055	0.4531
44.559	(111)	50 °C	1.5731	1.9102
51.930	(200)		0.4211	0.2365

the axes of the NTs. An increase in the temperature of the electrolyte leads to an increase in the intensity of the (111) peak with regard to the (200) peak as this occurs with increasing plate voltage.

Changes in the degree of texturing of the crystallites that made up the NTs due to variations in the conditions of synthesis were determined by studying texture coefficients TC_{hkl}. The values of the texture coefficients that exceed unity refer to the preferred orientation of crystallites in the corresponding planes (Table 7.5).

The obtained values of the texture coefficients TC_{hkl} show that the [111] direction dominates in the structure of the nickel NTs. With an increase in the potential and the synthesis temperature, the degree of texturing of the NTs increases.

To explain the transformations of the crystalline structure, specific features of the processes of electrochemical crystallization that occur during the precipitation of nickel should be considered. In the course of the formation of nickel NTs, the growth surfaces of the crystallites that make up the NTs present their (111) faces with the highest reticular density to the electrolyte. Due to the dependence of the crystallization pressure on the orientation of the predominantly growing close-packed faces, some of the adsorbed impurities becomes trapped in the precipitate, while the remainder of these impurities is primarily displaced in the direction perpendicular to the growth surface. The major impurity that can be

Table 7.6 Main structural parameters of Ni nanotubes synthesized at different temperature

Synthesis temperature, °C	Lattice parameter, Å		Coherent scattering region, nm		Crystallinity, %		δ, m²·1015	
	1,5 V	2 V	1,5 V	2 V	1,5 V	2 V	1,5 V	2 V
25	3.5179	3.5208	25.3	27.8	89.3	86.9	0.257	0,058
35	3.5196	3.5221	25.5	28.3	86.2	81.2	0.189	0.043
50	3.5203	3.5265	27.7	31.2	84.6	73.1	0.082	0.285

incorporated into the crystallites and form interstitial solid solutions is hydrogen, which is abundantly liberated at the cathode under the assumed conditions of electrodeposition. Particles of the incorporated impurities localized on the (111) plane play the role of active centers of the origination of monoatomic layers. Because of this, the next monoatomic layers on the surface of the NTs grow faster than the underlayers and, over time, crystallites appear elongated in the direction of the axis of the NTs.

An analysis of the dynamics of the lattice parameter a and the average crystallite size L, which was calculated using the Scherrer equation, has shown that, with an increase in the deposition potential and temperature, the lattice parameter and the average crystallite size increase (Table 7.6). It is in a good correlation with the results shown in the Sect. 3.2.

An increase in the deposition potential and/or the synthesis temperature results in the predominant growth of the (111) planes since the probability of particles of the codeposited impurities being incorporated into the crystallites; therefore, this leads to an increase in the lattice parameter (Fig. 7.7a). A part of the adsorbed impurities is displaced in the direction perpendicular to the (111) planes of the crystallites, which hampers the precipitation of the metal thereon. With an increase in the potential and/or the synthesis temperature, the rate of deposition rises, which favors the orientation of the crystallites along the axis of the NTs (Table 7.5) and leads to growth in the average crystallite size (Table 7.6).

7.3.4 *Influence of Diameters of Template Pore on Morphology and Structure of Nickel Nanotubes*

One of the interesting task is the change in the Ni NTs structure, and, accordingly, in the physical properties with variation of the characteristics of the template, in particular, the change in the pore diameters. Figure 7.8a shows SEM images of Ni NTs with different external diameters synthesized at 1.5 V.

Figure 7.8b shows that wall thickness monotonically rise with NTs external diameter increasing. At smallest diameter (100 nm) wall thickness is about 25 nm and the NSs are thin and brittle. This indicates on difficult to produce strength hollow nanostructures with diameter about 100 nm. On Fig. 7.9 there are the X-ray diffraction patterns of the samples under study, depending on the diameter of the NSs.

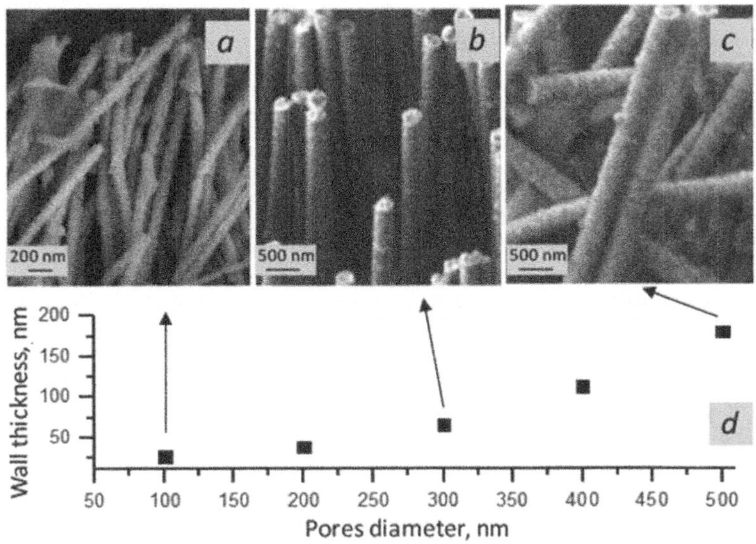

Fig. 7.8 SEM images of the Ni nanotubes synthesized in pores of PET template with diameters: (**a**) 100 nm; (**b**) 300 nm; (**c**) 500 nm; (**d**) dependence of Ni nanotubes wall thickness on pores diameter at constant deposition potential

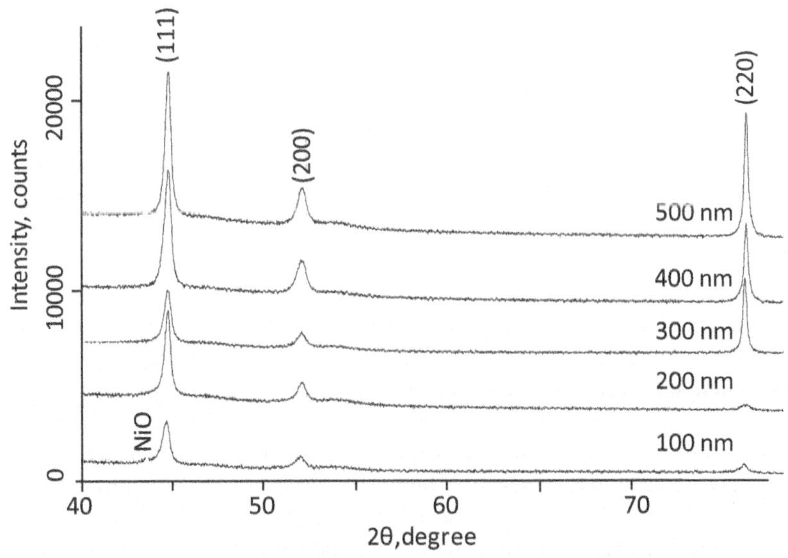

Fig. 7.9 XRD spectra of Ni nanotubes synthesized in pores of PET template with different diameters

Table 7.7 Texture coefficients of Ni nanotubes synthesized in pores of PET template with different diameters

	TC_{hkl}				
(hkl)	100 nm	200 nm	300 nm	400 nm	500 nm
(111)	2.311	2.156	1.151	1.031	1.297
(200)	0.365	0.642	0.456	0.675	0.667
(220)	0.324	0.202	1.393	1.294	1.036

Table 7.8 Main structural parameters of Ni nanotubes synthesized in pores of PET template with different diameters

Pores diameters, nm	Lattice parameter, Å	Coherent scattering region, nm	Cristallinity, %	δ, m2·1015
100	35.149	23.9	63.5	0.074
200	35.094	26.5	68.7	0.054
300	35.066	31.8	78.6	0.031
400	35.052	29.1	87.6	0.04
500	35.034	31.3	93.2	0.032

Ni NTs with diameter of 100 nm have a low-intensity peak characteristic of NiO which arise during process of NTs formation. The rapid rate of synthesis in small pores volume leads to the formation of amorphous inclusions and disorder regions in the structure, which are caused by oxygen ions coming into the crystal lattice during the synthesis. The content of the oxide phase in the structure does not exceed 6–8%. The degree of texturing of the crystallites in NTs with different diameters were determined by studying texture coefficients TC_{hkl}. The values are shown in Table 7.7.

An increase in the diameter of NTs leads to rise of the contribution of the texture plane (220) and, consequently, an increase in the degree of polycrystallinity of NTs. Main parameters of crystal structure are shown in the Table 7.8.

As can be seen from the presented data, the presence of oxide impurities and disorder regions leads to a low degree of crystallinity of NSs. Increasing the pore diameter at constant deposition potential leads to rise of synthesis time and causes the formation of a more ordered structure. Analyzing the FWDH and area of the diffraction peaks make it possible to estimate the contribution of various defects to the change in NTs structure. The broadening of the peaks with decreasing of Ni NTs external diameters take place due to microstrains in the structure, which are associated with the accumulation of dislocations, as well as the crushing of crystallites associated with crystallization processes. For example, for NTs with smallest diameter, a large number of different deformations and distortions of the crystal structure are characteristic, which is caused by the influence of various impurity inclusions (in particular, the oxide phase). An increase in the diameter of NSs leads to a decrease in microstresses and deformation in the lattice, however, for nanostructures with a diameter of 500 nm, there is an increase in distortion in the structure, which may be due to the appearance of additional texture orientations in nanotubes.

7.4 Magnetic Properties of Ni Nanotubes with Different Structure and Morphology

The main factors determining magnetic parameters as well as presence of magnetic anisotropy are magnetocrystalline anisotropy, dipole interaction, elastic stresses in NTs structure, and shape anisotropy. Nickel is a ferromagnetic with rather low coercivity and it's well known that the anisotropy is determined by shape anisotropy of Ni crystals. The easy magnetization axis of Ni corresponding to the (111) texture plane. Mostly, the direction [111] for our samples is predominant and it's orientation is along NTs axis. That is the reason of appearance of easy magnetic axis along the NTs.

As it was shown in paragraph 3, Ni NTs are polycrystalline and with presence of dislocation in the structure. The amount of dislocations and presence of impurities and amorphous inclusions that can have an effect on the magnetic properties, determined by anisotropic interactions in crystal, such as quantum spin-orbit interaction and magnetic dipole interaction (Table 7.9).

In this paragraph we discuss in details the changing of magnetic parameters and magnetic anisotropy raising with varying of structural and morphology parameters for series of the samples, analized in paragraph 3.

7.4.1 *Magnetic Properties of Ni Nanotubes with Different Length*

The values of the main magnetic characteristics should depend significantly on the NTs length and they change almost twofold during their growth. The loops measured at different directions of the magnetic field to the main axis of the NTs have different shape, which indicates the presence of the magnetic anisotropy in the samples: coercivity (H_C) and squareness ratio of hysteresis loop (M_r/M_S) for Ni NTs for parallel orientation are higher than the values for the perpendicular field direction (Fig. 7.10 and Table 7.9).

The dominant direction for the all samples of this series is [111] (Table 7.2). Those, the anisotropy of Ni NTs could be determined by shape anisotropy, but its contribution is varying as well as the role of crystalline anisotropy, which unfolds the magnetic moment perpendicular to the NTs axis. Magnetocrystalline anisotropy decreases with the stage of growth of NTs, cause the numbers of defects and amorphous inclusions, determined by decrease and cristallinity degree and dislocation density.

Thus, as far as more homogeneous the crystal structure of Ni NTs becomes, the magnetic anisotropy runs stronger, which is observed for samples (1)–(3). A further increase of the deposition time leads to the appearance of the first "caps", and subsequent their coalescence. When the "caps" are formed, several crystal growth

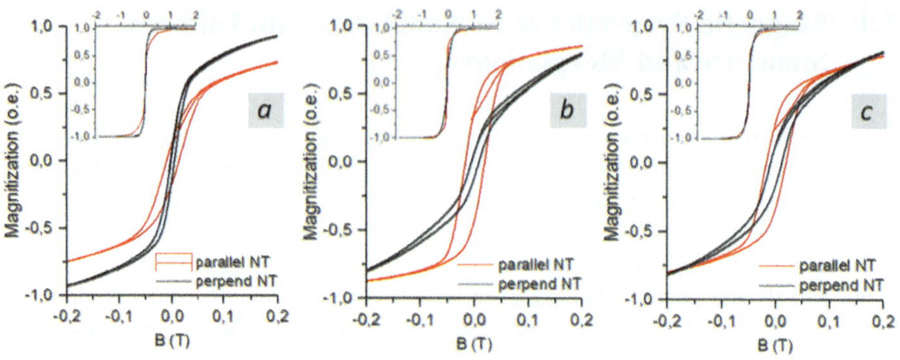

Fig. 7.10 Hysteresis loops of Ni nanotubes with different length: (**a**) 2.5 μm, (**b**) 7.5 μm, (**c**) 12.5 μm

Table 7.9 The main magnetic characteristics of Ni nanotubes with different length

	Main magnetic characteristics of nanotubes for field orientation			
	parallel to nanotubes axis		perpendicular to nanotubes axis	
Nanotubes length, μm	H_C, Oe	M_r/M_s	H_C, Oe	M_r/M_s
2.5	117	0.15	40	0.11
5	175	0.34	68	0.10
7.5	180	0.41	70	0.09
10	161	0.37	80	0.16
12.5	187	0.32	109	0.15

directions corresponding to the (111) and (200) texture planes occur, which entails an increase of the contribution of crystallographic anisotropy and decrease in the total magnetic anisotropy for samples (4)–(5).

7.4.2 Magnetic Parameters of Ni Nanotubes Synthesized at Different Deposition Potential

The hysteresis loops for samples, discussed in paragraph 3.2 are presented on Fig. 7.11. The main magnetic characteristics were determined and are presented in Table 7.10.

The study of hysteresis loops of a series of samples shows that the synthesized samples have the strong anisotropy. This is easily seen from the general view of hysteresis loops for a series of samples for different orientations of the magnetic field relative to the NTs axis. Coercivity and squareness ratio for a parallel orientation is nearly twice larger then for the perpendicular direction. Thus, the axis of easy magnetization is oriented along the NTs axis, but it does not coincide with it. The effect of crystalline anisotropy on magnetic properties of NTs has also been established. As the applied potential increases, the crystal structure of NTs

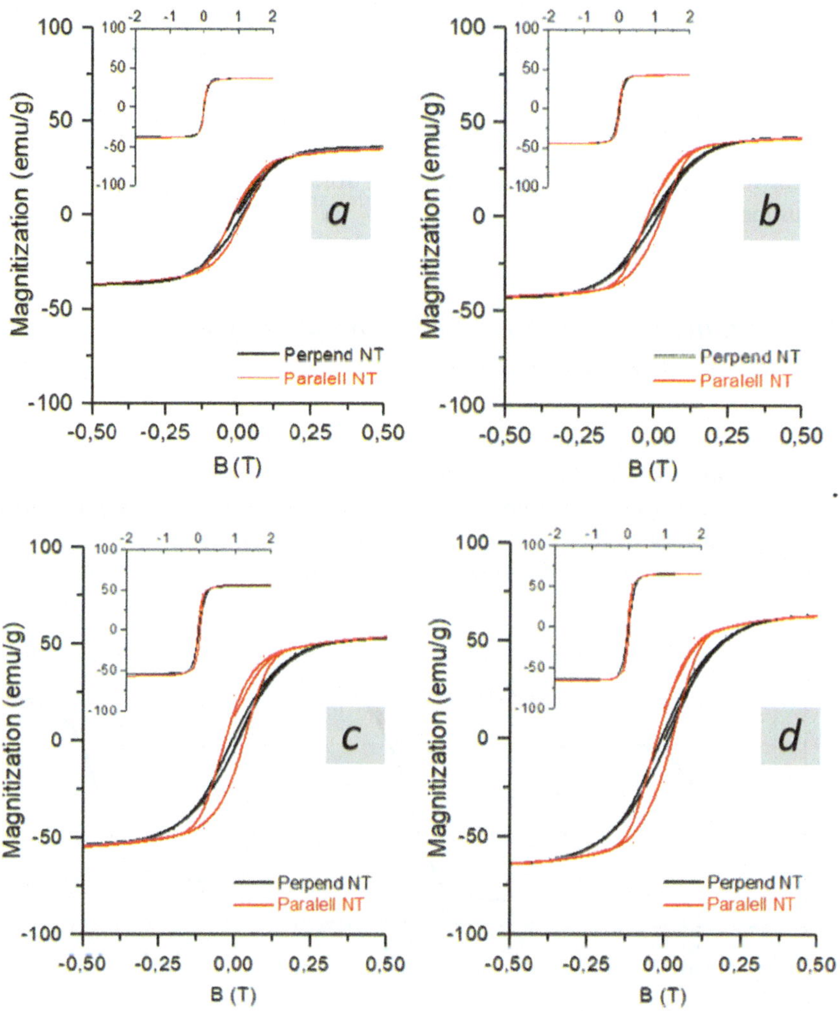

Fig. 7.11 Hysteresis loops of Ni nanotubes synthesized at different deposition potentials: (**a**) –1.25 V; (**b**) –1.5 V; (**c**) –1.75 V; (**d**) –2.0 V

Table 7.10 The main magnetic characteristics of Ni nanotubes synthesized at different potential

	Main magnetic characteristics of nanotubes for field orientation			
	parallel to nanotubes axis		perpendicular to nanotubes axis	
Potential of synthesis, V	H_C, Oe	M_r/M_s	H_C, Oe	M_r/M_s
1.25	240	0.18	165	0.12
1.5	275	0.22	150	0.08
1.75	310	0.28	160	0.10
2	290	0.26	160	0.09

changes. The Crystallographic planes and Miller indexes (111) is determined as the main one, but texture coefficient change with potential increase and the rise of this direction is determined. In this case, a slight increase in the magnetic parameters and an increase in the magnetic anisotropy can be observed. However, for 2.0 V applied potential, impurities are observed in the structure, which leads to a deterioration in magnetic characteristics and growth of the role of crystalline anisotropy, that unwraps magnetic moment.

7.4.3 *Magnetic Parameters of Ni Nanotubes Synthesized at Different Temperatures*

The hysteresis loops for samples, discussed in paragraph 3.3 are shown on Fig. 7.12 and the main magnetic characteristics are given in Table 7.11.

The strong magnetic anisotropy of Ni NTs arrays is observed: in the entire range of the synthesis parameters coercivity value for parallel orientation of the field relative to the NTs axis is several times higher than the values for the perpendicular.

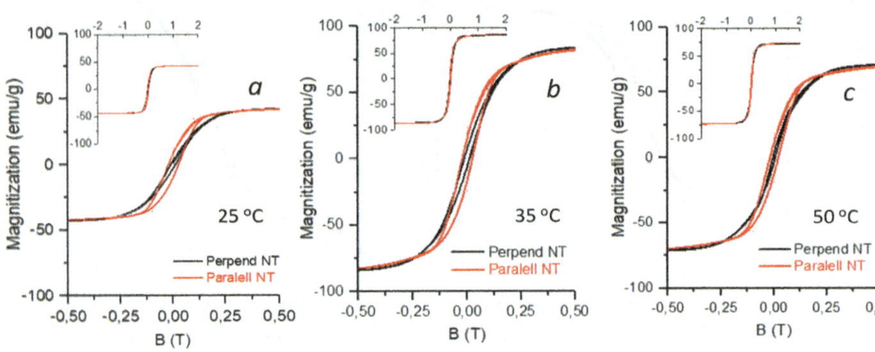

Fig. 7.12 Hysteresis loop of Ni nanotubes synthesized at potential 1.5 V at temperature: (**a**) 25 °C; (**b**) 35 °C; (**c**) 50 °C

Table 7.11 Magnetic parameters of Ni nanotubes synthesized at different temperatures

Synthesis temperature, °C	Voltage, V	Magnetic characteristics of nanotubes for field orientation			
		Parallel to nanotubes axis		Perpendicular to nanotubes axis	
		H_C, Oe	M_r/M_s	H_C, Oe	M_r/M_s
25	1.5	260	0.31	140	0.10
	2	280	0.28	150	0.09
35	1.5	280	0.25	150	0.12
	2	250	0.24	130	0.11
50	1.5	250	0.25	80	0.08
	2	320	0.29	250	0.13

Texture coefficient for all samples discussed in 3.3 have dominant orientation (111) that provides a strong anisotropy and its rising with temperature and potential. For the sample, prepared in condition 2 V and 50 °C, which has porous structure with thing walls and deteriorated structure the anisotropy decrease. For this sample magnetic parameters increase that corresponds to crystals growth (Table 7.6).

7.4.4 Magnetic Properties of Ni Nanotubes with Different Diameters

The hysteresis loops for samples, discussed in paragraph 3.4 are shown on Fig. 7.13 and the main magnetic characteristics are given in Table 7.12.

There is an interesting influence of pore diameter on magnetic properties of Ni NTs and magnetic anisotropy. Coercivity and squareness ratio arise for samples from 100 to 300 nm in diameter and fall down for 400 and 500 nm. Two factor play dominant role in magnetic properties evaluation. Accordingly to Tables 7.7 and 7.8 the growth of cristallinity degree, lattice parameter and crystals size with diameter

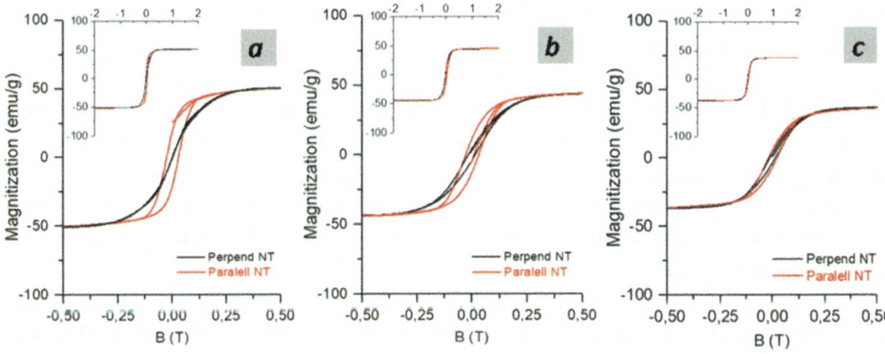

Fig. 7.13 Hysteresis loop of Ni nanotubes synthesized in pores of PET template with diameters: (**a**) 100 nm; (**b**) 300 nm; (**c**) 500 nm

Table 7.12 The main magnetic characteristics of Ni nanotubes synthesized in pores of PET template with different diameters

	Main magnetic characteristics of nanotubes for field orientation			
	Parallel to nanotubes axis		Perpendicular to nanotubes axis	
Pore diameter, nm	H_C, Oe	M_r/M_s	H_C, Oe	M_r/M_s
100	210	0.36	128	0.05
200	250	0,38	135	0,078
300	510	0,47	300	0,27
400	300	0,37	150	0,24
500	240	0,25	160	0,14

of NTs, and the redistribution of texture coefficients, which correspond to the main growth direction during the deposition process take place. As it was mentioned above, that the perfection of structure influence on shape anisotropy and presents of defects and stresses as well as impure determines the decrease of magnetization and changes in other magnetic parameters. Redistribution of picks on XRD-diagram, which means that the main NTs growth direction becomes not along to NTs (111) determines the decrease of magnetic anisotropy.

7.5 Conclusions

Structure, morphology and magnetic characteristics of nickel nanotubes synthesized in the pores of PET template with 100–500 nm pores diameters at different deposition time, temperature and potential were analyzed. On the base of time dependence study, mechanism of Ni nanotubes formation was established and evolution of structural parameters during nanotubes growth process was analyzed. It was shown that an increase in nanotubes length leads to an increase of crystallinity degree. The change of structural parameters is closely related to nanotubes electrodeposition stages: from the defective structure with the smallest crystallinity degree at first deposition stage to the polycrystalline low-defective structure at the moment of achievement of nanotube of total pore length.

Results of investigations of dependences of Ni nanotubes structural and morphological characteristic on temperature, potential and pore diameters shown that variation of these parameters leads to changes of structural perfection and wall thickness. It was demonstrated that these changes are strongly connected with deposition rate. Increasing of deposition rate and leads to rise of defects amount and oxide impurities with deterioration of structural properties and appearance of amorphous inclusions.

On the basis of the study of magnetic properties, the presence of magnetic anisotropy associated with the shape anisotropy Ni NTs was shown. By the detailed analysis of hysteresis loops dependences of magnetic characteristics on synthesis conditions were demonstrated. Changes of nickel nanotubes magnetic characteristics are caused of the morphology and structural parameters evolution, which are connected with deterioration of crystalline structure and appearance of amorphous inclusions.

Acknowledgements The authors acknowledge the support of the work in frames of H2020 – MSCA – RISE2017 – 778308 – SPINMULTIFILM Project, and Belarusian Foundation for Basic Research [project number Ф18М-080] and [project number Ф18КА3Г-001].

References

1. Akbarzadeh A, Samiei M, Davaran S (2012) Magnetic nanoparticles : preparation, physical properties, and applications in biomedicine. Nanoscale Res Lett 7:1–13
2. Yeszhanov AB, Mashentseva AA, Korolkov IV, Gorin YG, Kozlovskiy AL (2018) Copper nanotube composite membrane as a catalyst in Mannich reaction. Chem Pap 72:1–6
3. Korolkov IV, Kozlovskiy AL, Gorin YG (2018) Immobilization of carborane derivatives on Ni/Fe nanotubes for BNCT. J Nanopart Res, 20:Article ID 240
4. Kozlovskiy AL et al (2017) Comprehensive study of Ni nanotubes for bioapplications: from synthesis to payloads attaching. J Nanomater 2017:Article ID 3060972
5. Sander MS, Côté MJ, Gu W, Kile BM, Tripp CP (2004) Template-assisted fabrication of dense, aligned arrays of Titania nanotubes with well-controlled dimensions on substrates. Adv Mater 16:2052–2057
6. Graham LM, Cho S, Kim SK, Noked M, Lee SB (2014) Role of boric acid in nickel nanotube electrodeposition: a surface-directed growth mechanism. Chem Commun 50:527–529
7. Alnassar M, Alfadhel A, Ivanov YP, Kosel J (2015) Magnetoelectric polymer nanocomposite for flexible electronics. J Appl Phys 117:17D711
8. Shumskaya A, Kaniukov E, Kutuzau M, Kozlovskiy A, Zdorovets M (2017) Electrodeposited ferromagnetic nanotubes: structure and magnetic properties (pp 1–4)
9. Roose B, Ummadisingu A, Correa-baena J, Saliba M (2017) Nano Energy Spontaneous crystal coalescence enables highly efficient perovskite solar cells. Nano Energy 39:24–29
10. Jovanović S, Spreitzer M, Tramšek M, Trontelj Z, Suvorov D (2014) Effect of oleic acid concentration on the physicochemical properties of cobalt ferrite nanoparticles. J Phys Chem C 118:13844–13856
11. Shumskaya AE et al (2017) Template synthesis and magnetic characterization of FeNi nanotubes. Prog Electromagn Res C 75:23–30
12. Shumskaya AE, Kaniukov EY, Kozlovskiy AL, Zdorovets MV, Rusakov VS, Kadyrzhanov KK (2017) Structure and physical properties of Iron nanotubes obtained by template synthesis. Phys Solid State 59:784–790
13. Martín JI et al (2002) Fabrication and magnetic properties of arrays of amorphous and polycrystalline ferromagnetic nanowires obtained by electron beam lithography. J Magn Magn Mater 249:156–162
14. Barth S et al (2009) Studies on surface facets and chemical composition of vapor grown one-dimensional magnetite nanostructures. Cryst Growth Des 9:1077–1081
15. Morber JR et al (2006) PLD-assisted VLS growth of aligned ferrite nanorods, nanowires, and nanobelts-synthesis, and properties. J Phys Chem B 110:21672–21679
16. Yarmolich M, Kalanda N, Demyanov S, Fedotova J, Bayev V, Sobolev NA (2016) Charge ordering and magnetic properties in nanosized $Sr_2FeMoO_{6-\delta}$ powders. Phys Status Solidi 253:2160–2166
17. Liu Z, Zhang Q, Shi G, Li Y, Wang H (2011) Solvothermal synthesis and magneto-optical properties of Zn1−xNixO hierarchical microspheres. J Magn Magn Mater 323:1022–1026
18. Hua Z et al (2006) Metal nanotubes prepared by a sol–gel method followed by a hydrogen reduction procedure. Nanotechnology 17:5106–5110
19. Zhou D, Wang T, Zhu MG, Guo ZH, Li W, Li FS (2011) Magnetic interaction in FeCo alloy nanotube array. J Magnet 16:413–416
20. Vivas LG, Ivanov YP, Trabada DG, Proenca MP, Chubykalo-Fesenko O, Vázquez M (2013) Magnetic properties of co nanopillar arrays prepared from alumina templates. Nanotechnology 24:105703
21. Martin CR (1994) Nanomaterials: a membrane-based synthetic approach. Science 266:1961–1966
22. Haehnel V et al (2010) Towards smooth and pure iron nanowires grown by electrodeposition in self-organized alumina membranes. Acta Mater 58:2330–2337

23. Giuliani J, Monton C (2017) Template-assisted electrodeposition of Ni and Ni/Au nanowires on planar and curved substrates. Nanotechnology 29:075301
24. Gehlawat D, Chauhan RP (2014) Swift heavy ions induced variation in the electronic transport through Cu nanowires. Mater Chem Phys 145:60–67
25. Samykano M, Mohan R, Aravamudhan S (2014) Morphology and crystallographic characterization of nickel nanowires—influence of magnetic field and current density during. Synth J Nanotechnol Eng Med 5:021005
26. Kalska-Szostko B, Wykowska U, Satuła D (2015) Magnetic nanowires (Fe, Fe-Co, Fe-Ni) – magnetic moment reorientation in respect of wires composition. Nukleonika 60:63–67
27. Kaniukov E, Shumskaya A, Yakimchuk D, Kozlovskiy A, Ibrayeva A, Zdorovets M (2017) Characterization of pet track membrane parameters. In: NANO 2016: Nanophysics, nanomaterials, interface studies, and applications, 2017. Springer, Cham, pp 79–91
28. Cao H et al (2006) Generation and growth mechanism of metal (Fe, Co, Ni) nanotube arrays. ChemPhysChem 7:1500–1504
29. Shimanovich DL, Vorobjova AI, Tishkevich DI, Trukhanov AV, Zdorovets MV, Kozlovskiy AL (2018) Preparation and morphology-dependent wettability of porous alumina membranes. Beilstein J Nanotechnol 9:1423–1436
30. Yakimchuk D et al (2018) Silver nanostructures evolution in porous SiO2/p-Si matrices for wide wavelength surface-enhanced Raman scattering applications. MRS Commun 8:95–99
31. Kaur A, Chauhan RP (2014) Effect of gamma irradiation on electrical and structural properties of Zn nanowires. Radiat Phys Chem 100:59–64
32. Kozlovskiy AL, Shlimas DI, Shumskaya AE, Kaniukov EY, Zdorovets MV, Kadyrzhanov KK (2017) Influence of electrodeposition parameters on structural and morphological features of Ni nanotubes. Phys Met Metallogr 118:164–169
33. Yoo B, Xiao F, Bozhilov KN, Herman J, Ryan MA, Myung NV (2007) Electrodeposition of thermoelectric superlattice nanowires. Adv Mater 19:296–299
34. Motoyama M, Fukunaka Y, Sakka T, Ogata YH (2007) Initial stages of electrodeposition of metal nanowires in nanoporous templates. Electrochim Acta 53:205–212

Chapter 8
First- and Second Order Light Scattering Processes in Biological Photonic Nanostructures

Géza I. Márk, Krisztián Kertész, Gábor Piszter, Zsolt Bálint, and László P. Biró

Abstract The colors of various butterflies often originate from photonic nanostructures, found in the scales covering their wings. Such colors are called structural colors. The color generating scales are composed of a nanostructured chitinous material containing air voids, which causes the structural colors through light interference. We performed optical spectrum simulations utilizing full 3D Maxwell equation calculations on model structures to reveal the connection between the 3D structure and the optical spectrum. Our simulations showed that different scattering processes determine the spectrum in different wavelength ranges. For large wavelengths (>350 nm) the optical reflection can be well described by a corresponding effective multilayer model and the peak positions are well represented by a simple first Born approximation. One has to include second order scattering processes inside the layers, however, in order to correctly reproduce the small wavelength side of the spectrum (<350 nm). This means that such details of structure, as the shape of the air voids determine the small wavelength spectrum.

Keywords Photonic crystals · Bioinspiration · First born approximation · Ewald sphere

---

G. I. Márk (✉) · K. Kertész · G. Piszter · L. P. Biró
Institute of Technical Physics and Materials Science, Centre for Energy Research, Budapest, Hungary
e-mail: mark@mfa.kfki.hu

Z. Bálint
Hungarian Natural History Museum, Budapest, Hungary

A. Maffucci, S. A. Maksimenko (eds.), *Fundamental and Applied Nano-Electromagnetics II*, NATO Science for Peace and Security Series B: Physics and Biophysics, https://doi.org/10.1007/978-94-024-1687-9_8

8.1 Introduction

The colors of animals [1] originate from pigments and photonic nanostructures, often from a complex interplay between them. Even when the optical properties of the materials building up photonic nanostructures do not show significant wavelength dependence in bulk form, the reflection function of the nanostructure can still show wavelength dependence. Colors originating from photonic nanostructures are called structural colors, and their most spectacular examples are seen on insects, beetles and butterflies. The most prominent optical phenomenon seen in butterfly wings is the so-called iridescence, [2] when the hue of the butterfly wing changes with the angle of observation and/or illumination. Butterfly wings are covered by scales, which are flat sacs of dried cuticle. The spatial dimension of a wing scale is limited, its thickness is generally only 1–2 μm and its lateral extension is 50–100 μm. While the lower – closer to the wing membrane – side is generally structureless, the upper side may have a complicated structure. The inside of the scale is often filled by nanostructured chitinous material, which causes [3] the structural colors by constructive and destructive interference of the electromagnetic waves scattered on these microstructures or nanoarchitectures. This nanostructured chitinous material is a nanocomposite built of mostly two basic components with different refraction indices [4, 5], air $n_{air} = 1$ and chitin, $n_{chitin} = 1.56$. The air is present in the system in the form of air voids with sizes and distances such that the reflection spectrum of the scale has peaks in the visible wavelength range, which is perceived as a structural color [6].

Butterfly scale microstructures can be classified by a number of ways, [vii] a detailed classification from the point of view of biology was given by Ghiradella [7, 8]. One of the simplest classification of the nanoarchitectures found in butterfly scales from a physicist's point of view is whether it has a structure in the length scale of the wavelength of visible light in one, two, or all the three dimensions (1D, 2D, and 3D structures). Optical properties of butterfly wing scales were first modeled by the theoretical concept of photonic crystal, [9] which, in its most general definition is a graded-refraction-index material, where the refractive index is a periodic function of the displacement in one-, two-, or three dimensions, giving rise to a stop band. The stop band, or photonic band gap, an analogue of the forbidden band in solid state physics, is a wavelength range where light can not travel through the crystal, hence it is reflected. Biological materials, however, always have certain kind of disorder and the disorder varies on a large scale among different butterfly species. Crystalline order, however, is not a necessary condition of the existence of photonic band gap, amorphous materials can also have [10] a photonic band gap, if a short-range order is present. Metallic-like reflectance, i.e., emergence angle selection is related to the photon momentum conservation, which follows from total or partial translational invariance in the diffusing surface structure. Hence the metallic-like reflectance is associated with highly correlated structures [2, 11]. It is remarkable, however, that not only shiny, but also matt appearance can be produced by scale nanostructures [12]. Matt appearance seems to always involve some form of long-range disorder. Both the shiny reflectance and the diffuse reflectance have specific biological function for the individual. Shiny reflectance is used for signaling, which is crucial for sexual communication, therefore the color is under sexual selection. The matt color has a cryptic function serving as a generalized camouflage function, thus affected by natural selection [13].

Most important tools to analyze the nanostructure of butterfly wing scales are scanning electron microscopy (SEM) and transmission electron microscopy (TEM). Top-view SEM images provide information, however, about the topmost layer only and the sample preparation (deposition of sputtered gold) may distort the appearance of the nanostructures on the images. Cross-sectional TEM images do give access to the in-depth structure of the scales, but the precise direction of the cut is generally not known—an uncertainty of 5–10° is common— and the samples are sometimes distorted during the preparation process. Both methods give basically 2D information and it is a non trivial task to determine the real 3D structure of the scale based on SEM and TEM images. A careful correlated examination of the SEM and TEM images [12] has to be carried out in order to obtain the full 3D structure.

Computational electrodynamics [14] can predict the optical properties of a material of known three-dimensional structure with high accuracy, utilizing frequency domain methods, such as the Finite Element Method (FEM), or time domain methods, such as the Finite-Difference Time-Domain Method (FDTD). These computations give exact results for perfectly ordered structures, but do have difficulties in handling randomness, because of the large computational demand. Butterfly wing scales, however, have a small refractive index contrast, small thickness, and low absorption. Hence it is possible to approximate their optical properties by simpler, but more intuitive methods, such as first order scattering methods.

In this Chapter we demonstrate the applicability and limitations of the first order scattering method for approximate calculation of the optical properties of butterfly wing scales. The organization of this Chapter is as follows. In Sect. 8.2 we review the basic formulas of the first order scattering method and demonstrate it on examples utilizing the Ewald sphere construction. In. Sect. 8.3 we investigate in detail, how the randomness, which is inherently present in the nanoarchitectures of biological origin, influences the light scattering process. By systematic analysis of different distortions of a perfect lattice we identify the signatures of disorder in the Fourier power spectrum. In Sect. 8.4. we apply these principles to real butterfly photonic nanostructures. Section 8.5 contains a comparison of the reflection spectrum calculated by full 3D electrodynamics with the first order approximation. Here we introduce the concept of second order scattering. Section 8.6 is devoted to the conclusions.

8.2 The First Born Approximation and the Ewald Sphere Method

The first order scattering intensity, according to basic scattering theory, [15] is given by

$$I\left(\vec{k};\vec{k_0}\right)=\int\rho\left(\vec{r}\right)\exp\left(i\frac{\vec{q}}{n_{eff}}\vec{r}\right)d^D\vec{r},\tag{8.1}$$

where $\rho\left(\vec{r}\right)$ is the density of the scattering centers, $\vec{k_0}$ is the wave vector of the incoming light, $\vec{k}$ is the wave vector of the outgoing (transmitted or reflected) light, (the magnitude of the two wave vectors is equal in case of elastic scattering), $\vec{q}=\vec{k}-\vec{k_0}$ is the scattering vector, and n_{eff} is the effective refractive index of the medium. The integral is a D dimensional integral, D = 3 in 3D calculations, but it is D = 2 in the 2D examples demonstrated in this section. This exponential integral has the same form as a Fourier integral and this is the basis of the so called Ewald construction, demonstrated in Fig. 8.1 which is widely used in the theory of elastic X-ray scattering. Figure 8.1a shows a hypothetic 2D periodic $\rho(x, y)$ distribution of the scattering centres and Fig. 8.1b its Fourier transform $\tilde{\rho}\left(k_x, k_y\right)$. The $\rho(x, y)$ is periodic, hence $\tilde{\rho}\left(k_x, k_y\right)$ consists of discrete points (Dirac deltas) – which become square pixels in the quantized numerical representation (Discrete Fourier Transform, DFT) of Fig. 8.1b. The $\rho(x, y)$ is, however, not harmonic (not a

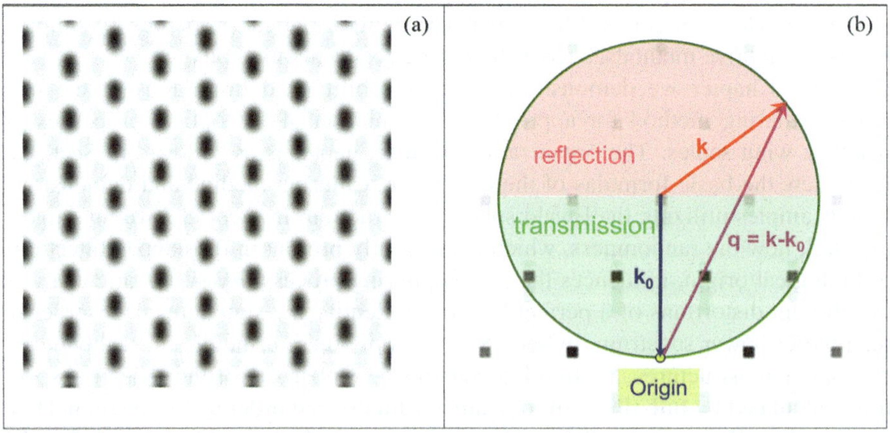

Fig. 8.1 Demonstration of the Ewald construction in two dimensions. (**a**) A simulated triangular periodic lattice (sum of three anharmonic plane waves in 120^0 directions). (**b**) Fourier transform of the lattice in (**a**) and the Ewald construction. $\vec{k_0}$ is the wave vector of the incoming light, $\vec{k}$ is the wave vector of the outgoing (transmitted or reflected) light, (the magnitude of the two wave vectors is equal in case of elastic scattering), $\vec{q}=\vec{k}-\vec{k_0}$ is the scattering vector

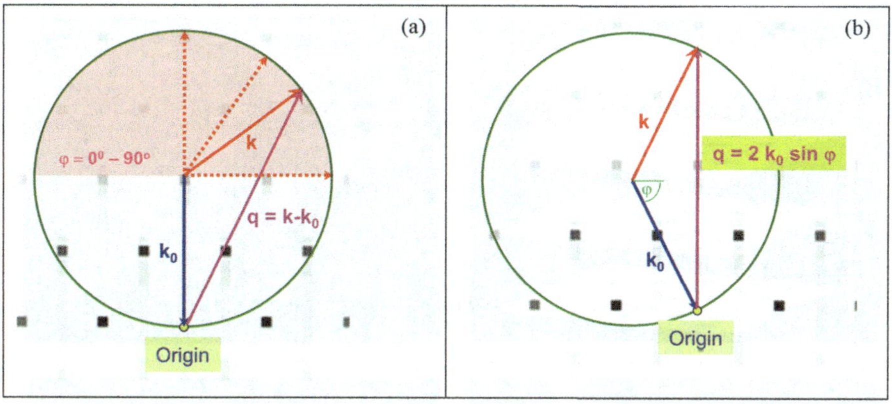

Fig. 8.2 Ewald construction for different cases. (**a**) When the angle of the outgoing light changes between 0–90 (from perpendicular scattering to normal backscattering), the length of the scattering vector changes from $q_{min} = \sqrt{2}k_0$ to $q_{max} = 2k_0$. (**b**) Specular (mirror) reflection. The length of the scattering vector depends on the incoming (=outgoing) angle as $q = 2k_0 \sin \varphi$

sinusoidal function), hence several harmonics ("overtones") are seen in $\tilde{\rho}\left(k_x, k_y\right)$. If the sample is illuminated with a monocromatic plane wave with wave vector $\overrightarrow{k_0}$ and we detect the light scattered in different directions, then the tip of the wave vector $\overrightarrow{k}$ of the scattered light moves along a sphere (a circle in 2D), because the scattering is elastic, hence $|k| = |k_0| = {2\pi}/{\lambda}$. This is called the Ewald sphere (circle). The scattered light intensity $I\left(\overrightarrow{k}\,;\,\overrightarrow{k_0}\right)$ in different directions is given by the value of the $\tilde{\rho}\left(k_x, k_y\right)$ along the Ewald circle. If the incoming light is white light, then different colors are scattered in different directions, as shown on Fig. 8.2a. Figure 8.2b shows the case of the specular (mirror) reflection.

When $\rho(x, y)$ is periodic, then the scattering occurs only in a few directions. This is demonstrated on two three dimensional (3D) examples of Fig. 8.3. In the first example, Fig. 8.3a the lateral arrangement of the scattering centers is a harmonic triangular function and its Fourier transform, Fig. 8.3b is a hexagon of 6 dirac delta peaks. In Fig. 8.3a an infinite number of such layers is stacked in the vertical direction, and the Fourier transform Fig. 8.3b contains also an infinite number of layers, because of the infinitesimal thickness of the layers in Fig. 8.3a. In the second example, Fig. 8.3c the real space function is harmonic in all directions, hence its Fourier transform, Fig. 8.3d is composed only of a finite number of Dirac delta peaks: two hexagons shifted in the z directions. Figure 8.3e shows again these two hexagons together with the Ewald sphere. If we illuminate this sample with a white light in the perpendicular direction there will be a wavelength and direction selection. If we vary the diameter of the Ewald sphere (different colors) it only touches the upper hexagon only for a specific diameter (color) and the scattering

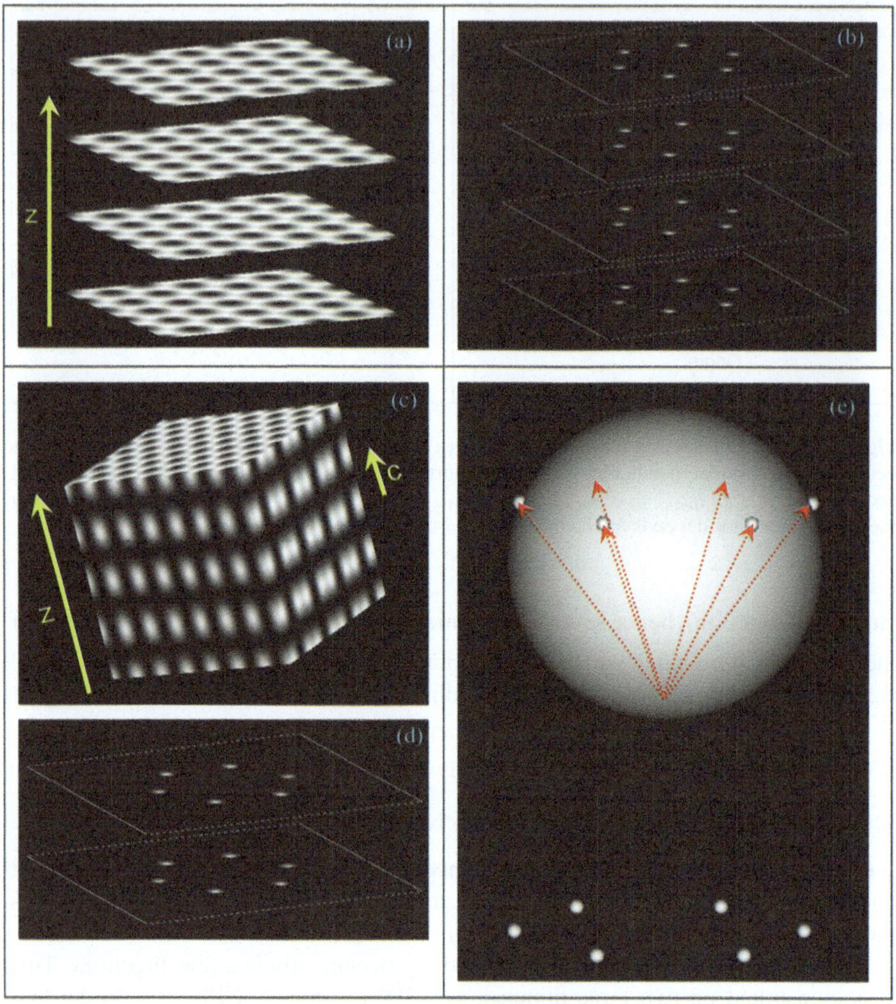

Fig. 8.3 Three dimensional Ewald construction for different cases. (**a**) Layers of harmonic triangular lattices stacked in the vertical (z) direction. The figure shows only 4 layers of the infinite stack. (**b**) Fourier transform of (**a**). It is composed of infinite number of layers having 6 points in a hexagonal orientation. Only 4 layers are shown. (**c**) A $\rho(x, y, z)$ function harmonic in all three directions. The figure shows only a rectangular volume cut from of the infinite function. (**d**) Fourier transform of (**c**). It is composed only of two hexagons. (**e**) 3D Ewald construction. Backscattering occurs only for one color and six angles determinded by the periodicities of $\rho(x, y, z)$

directions of this color are determined by the touching points (denoted by the six red vectors in Fig. 8.3e). It is interesting that there is no forward scattering (transmittance) in this system because the Ewald sphere does not cross the lower hexagon whatever the diameter is.

8.3 Order-Disorder Effects in the First Order Scattering

The model geometries analyzed in Sect. 8.2. were perfect infinite crystal structures, possessing a perfect translational symmetry. If a structure has a perfect long range order, its Fourier transform is composed of distinct peaks. This situation, is seen in X-ray scattering images in solid state physics, because atomic or molecular crystals can have very regular structures. Photonic nanostructures found in biological systems differ from this ideal situation in two aspects:

1. The structure has a finite size, especially its thickness (z direction) is limited, only several microns. Cutting an infinite structure to a finite size corresponds to a convolution of the Fourier transform of the infinite structure with the Fourier transform of the window function (e.g. sin x/x for the case of a rectangular window).
2. Biological systems always contain disorder. If the real space structure is not as regular as crystals in solid state physics, the Fourier transform is not any more composed of Dirac delta like peaks, but it has a complicated, continuous distribution in reciprocal space.

First we analyze the effect of the finite size. Figure 8.4 shows the effect of the spatial confinement of the system, it is composed only of four layers, but the layers are assumed to be infinite in the x direction. As seen on Fig. 8.4b and d, the Fourier peaks are broadened in the y direction and satellite peaks appear between the original peaks. For the case of n layers there are n−1 zero nodes and n−2 satellite peaks. Figure 8.4c is a function constructed as a sum of four Gaussians placed on an equidistant grid and Fig. 8.4d shows its Fourier power spectrum, which is a convolution of a Gaussian with a Sin x/x function. We can see three zero nodes and two maxima. This process is analyzed in detail in [16].

Next we turn to the investigation of the randomness. The upper left subimage of Fig. 8.5 shows a rectangular section of a perfectly regular and infinite 2D square lattice and its Fourier power spectrum. The Fourier image is a square lattice of distinct peaks. The intensity of the peaks is decreasing with the distance from the origin, this decrease is faster if the objects placed on the lattice points of the real space lattice are larger. As we now from the theory of X-ray scattering, if N objects are placed in the points of a direct space lattice, the total scattering of the system is given as the product of the lattice factor (Fourier transform of the lattice) and the so called atomic form factor (Fourier transform of the repeated pattern). The Fourier image is composed of distinct peaks in the horizontal (vertical) direction if the lattice has a long range periodicity in the x (y) direction.

We introduce two different randomness into this "perfect" picture:

- We preserve the perfect periodicity inside the layers, but the lattice layers are randomly displaced horizontally. This is shown in the upper row of Fig. 8.5, for increasing displacements. We can see that the Fourier images are composed of *vertical* lines. This is because the random displacement of the layers destroys the

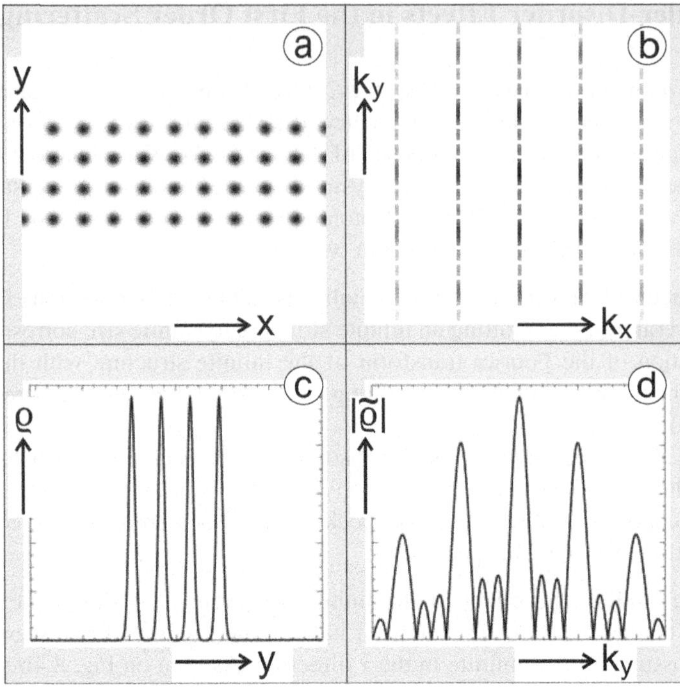

Fig. 8.4 Effect of the finite number of layers on the Fourier power spectrum. (left) Real space images. (right) Reciprocal space images. (**a**) and (**b**) Four layers of a perfect rectangular lattice. The structure is infinite in the x direction, only a finite section is shown. (**c**) Vertical density distribution $\rho(x = 0, y)$ in the central column of (**a**). (**d**) Vertical Fourier power density distribution $\tilde{\rho}\left(k_x = 0, k_y\right)$ in the central column of (**b**). Dimensionless units are used in all subimages. See the text for details

perfect translational symmetry in y. The image is still composed of discrete lines, because the layers still have a perfect translational order in the x direction.

- The leftmost column of Fig. 8.5 shows the effect of the disorder present in the layers themselves. This time a random horizontal displacement with a Gaussian distribution was applied to the points of one layer and this layer was repeated infinitely in the y direction. All of the layers have the same structure this time. The Fourier image shows distinct *horizontal* lines. Indeed, each reciprocal lattice point is broadened in the x direction because of the random x displacement of the points, but the translational order is present in the y direction because the layers have an identical structure.
- The inner subimages of Fig. 8.5 display the effect of applying the two kind of randomizations together. We have here both vertical and horizontal broadening of the Fourier peaks at hand.

Another important kind of random structure is the polycrystalline arrangement, this is shown on Fig. 8.6. First we created a perfect harmonic triangular lattice

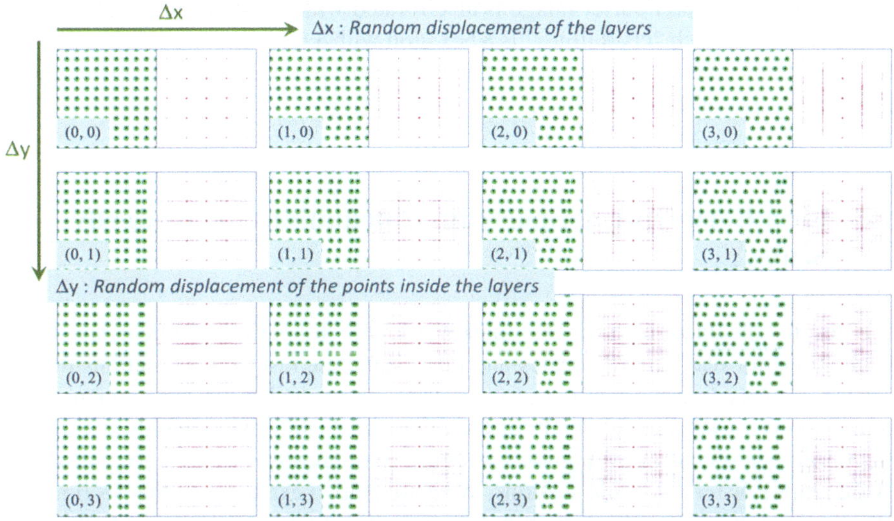

Fig. 8.5 Effect of the randomness on the Fourier power spectrum. The left panes of each subimages show a real space lattice and the right panes show its Fourier power spectrum. The real space lattices are infinite in the calculations, but only a finite window is shown. The subimages are arranged in a matrix, where the columns and the rows corresponds to two different randomization parameters. The upper left subimage is a perfect square lattice. The upper row shows the effect of the random displacement of the layers with increasing displacement from left to right. The leftmost column shows the effect of the random displacement of the points inside the layers with displacement increasing from up to bottom. The inside images are for the combination of both kind of randomization – the reciprocal lattice peaks are broadened in both the vertical and horizontal direction

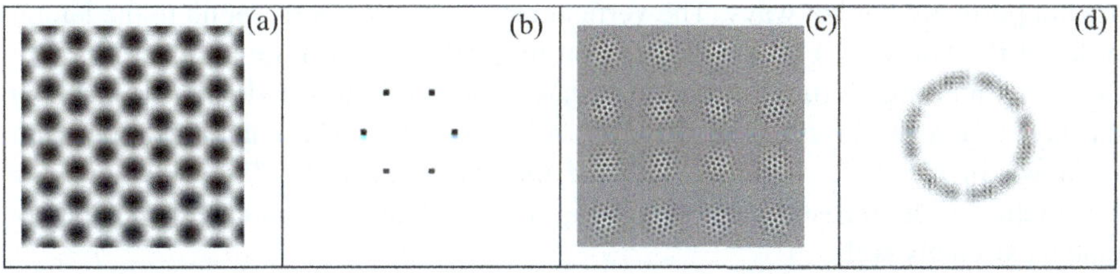

Fig. 8.6 Fourier transform of a polycrystalline lattice. (**a**) A harmonic triangular lattice. (**b**) Fourier transform of (**a**) – six points at the vertices of a hexagon. (**c**) Polycrystalline lattice. (**d**) Fourier transform of the polycrystalline lattice

(Fig. 8.6a), its Fourier transform (Fig 8.6b) contains only six points at the vertices of a hexagon. Next we cut (by a 2D Gaussian window function) a finite part of the crystal lattice and placed it to the points of a rectangular superlattice. Then we applied random rotations to each of these "crystallites". The resulting real space polycrystalline lattice if shown on Fig. 8.6c and its Fourier transform on Fig. 8.6d. The points building up this Fourier image are arranged along a circle. This is because of the rotation invariance and additivity of the Fourier transform: if we rotate the real space image, then the Fourier image is also rotated with the same angle. Hence the Fourier image of each "crystallites" is a hexagon with a random rotation. The sum of the rotated hexagons gives the circle. The radial broadening of the circle is because of the finite size of the crystallites.

8.4 First Born Optical Spectrum Calculations for Butterfly Scales

The results obtained in Sect. 8.3 makes us able to perform optical spectrum calculations for real butterfly scales and understand the results. Calculations in this Section are based on the method of [17]. They first calculate the two-dimensional Fourier power spectrum (2D FFT) of the binarized TEM image then they estimate the reflectance spectrum by angular averaging of the power spectrum for the whole $[0, 2\pi)$ range or for $^{\pi}/_{6}$ wide sections. We analyze two examples in this section: (i) *Albulina metallica* butterfly, which has a randomized layer structure on both sides and (ii) *Cyanophrys remus* butterfly, which has a polycrystalline structure on its ventral side.

The dorsal side of the Albulina butterfly is metallic blue, the ventral side is silvery green (cf. Fig. 8.7a). Figure 8.7a also shows the cross sectional TEM images, Fig. 8.7b is the binarized TEM image, and Fig. 8.7c is its Fourier power spectrum. The vertical structure A^{-}–O–A^{+} corresponds to the layer structure. Note that there are two minima and one maximum between each pair of main peaks, which corresponds to the presence of three layers, as it was explained in Sect. 8.3, see Fig. 8.4. The peaks at A^{-}, A^{+} are broadened horizontally because the layers are not perfectly flat but wavy. The vertical bars at $B-$,$B+$ correspond to the lateral order of the holes, cf. Fig. 8.5. These Fourier peaks have a finite thickness in both the horizontal and vertical direction because the holes are randomly arranged in the layers and the layers have different microstructure. The calculated spectrum is shown in Fig. 8.8c, together with the measured spectrum (Fig. 8.8b) and the schematics of the measurement setup (Fig. 8.8a). Measured and calculated spectra match extremely well.

The inset of Fig. 8.7c shows the Fourier image of an idealized structure, where the lateral order is perfect in the layers, but the layers are randomly displaced in the lateral direction. This corresponds to the model shown in Fig. 8.5. The spectrum of this ideal structure can be calculated easily by the Ewald construction and it is shown by the dotted line in Fig. 8.8c. This theoretical curve matches the trend of the peak positions calculated from the Fourier transform of the TEM section. Details of this calculation are given in Ref. [16].

Our example for the polycrystalline wing scale structure is the ventral side of the *Cyanophrys remus* butterfly. The ventral side of this butterfly (cf. Fig. 8.9a) has a dull, matt green color. Surprisingly, optical microscopy (cf. Fig. 8.9b) reveals (in reflection) that the overall matt, pea-green coloration of the ventral wings arises from the mixing of very tiny, bright, blue, green, and yellow spots

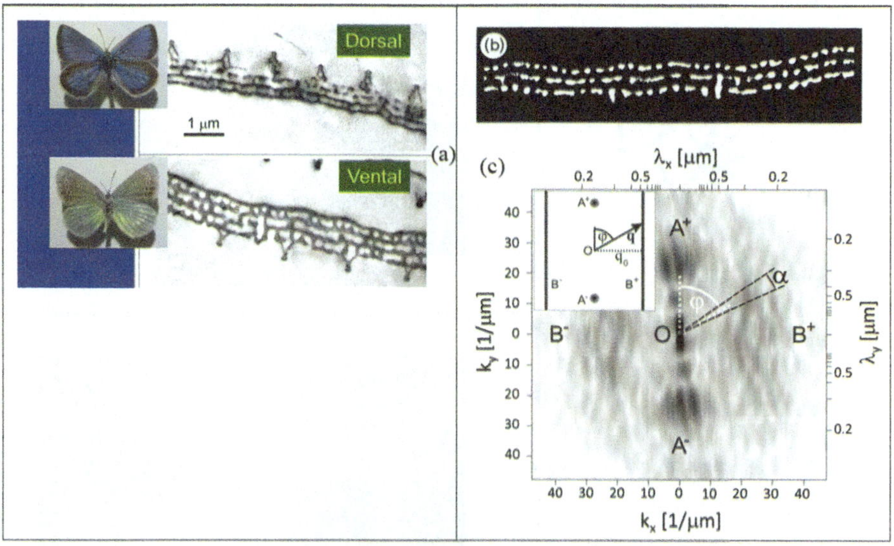

Fig. 8.7 Fourier analysis of the TEM images of the *Albulina metallica* butterfly. (**a**) Optical image of the dorsal and ventral side; cross sectional TEM images. (**b**) Processed TEM image of a ventral scale of *Albulina metallica*. A background correction and a binarizing filter was applied. (**c**) Fourier power spectrum of the processed TEM image. White corresponds to zero intensity and black to maximum intensity. Wave number (wavelength) scale is shown on the lower and left (upper and right) axis, respectively. The lines illustrate the procedure of the calculation of the backscattered spectrum from the Fourier image. The inset shows the Fourier image and the procedure of the spectrum calculation for a model image

randomly distributed over the scale. The dominant color and the intensity of the bright localized reflectors show a significant sensitivity to the observation and illumination directions, indicative of a local structural coloration. The SEM and TEM investigation of the wings [7] revealed scales with a rather peculiar structure: consisting of small grains with typical diameters in the 5–10 μm range. These grains show a regular structure, the regularity of the structure within one single grain is convincingly revealed by the cross-sectional TEM images in Fig. 8.9c. The grains act as individual photonic crystallites and, depending on the orientation of the incident light and the observation angle, give rise to the bright color spots detected in optical microscopy (cf. Fig. 8.9b). The summation of the colors and the random orientation of the grains may be responsible for the matt, pea-green coloration of the ventral scales.

As we explained in Sect. 8.3, the Fourier transform of a polycrystalline structure (cf. Fig. 8.6) is a ring like structure (a sphere in 3D), indicative of the presence of a radial order and no angular order. This Fourier image is similar to those obtained for photonic glasses and as was shown in the detailed analysis of Mainwald et al. [18], this structure leads to non-iridescent colors.

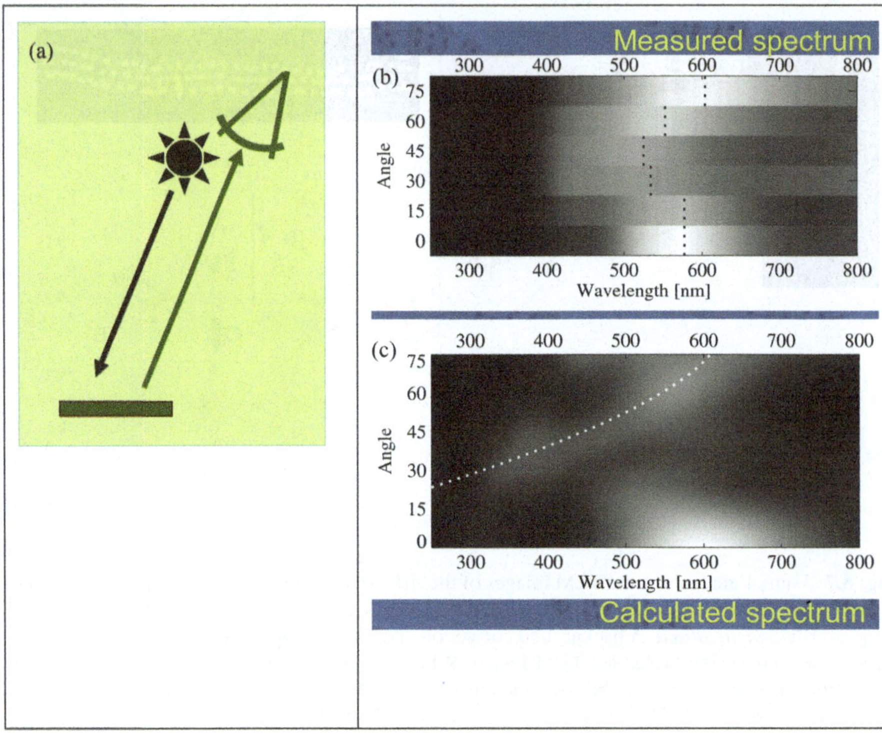

Fig. 8.8 Calculation of the backscattered spectrum and comparison to the measurement. (**a**) The backscattered measurement configuration. The illumination and the detector are placed at the same angle. (**b**) Reflectance spectrum of the *Albulina metallica* ventral (green) side recorded at backscattering arrangement at several angles. All spectra were measured relative to a diffuse white standard. Black corresponds to zero intensity and white to maximum intensity. A nonlinear gray scale is used to facilitate presentation. Black dotted lines show the wavelengths with maximum intensity for each angle. (**c**) Theoretical backscattered reflectance spectra calculated from the Fourier power spectrum shown on Fig. 8.7c. The dotted line shows the spectrum calculated for the model in Fig. 8.5

8.5 Signatures of Higher Order Scattering in the Optical Spectrum

In order to assess the validity of the first order approximation for the butterfly scales, we compared the first Born calculated spectra with an "exact" full 3D Finite Element (FEM) electrodynamical calculation (https://www.comsol.com/wave-optics-module). The model structure consists of four layers. Each layer is an infinite square lattice of dielectric spheres. The lateral lattice constant is d = 300 nm and the layer distance is a = 250 nm. Figure 8.10a shows the unit cell used in the calculation. Free space boundary conditions were used at the +z and –z boundaries and Floquet periodic boundary conditions were applied in the lateral, −x,+x,−y,+y directions.

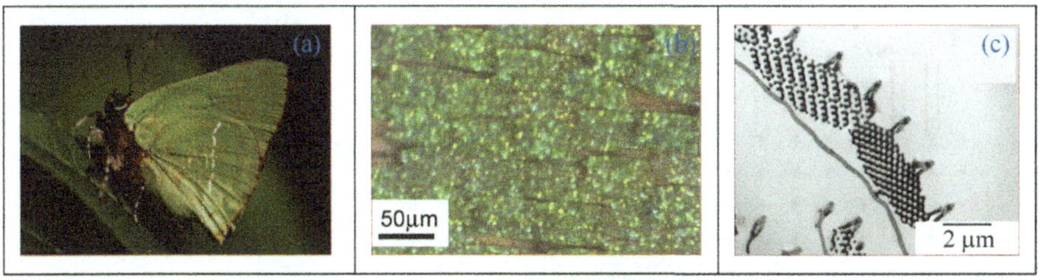

Fig. 8.9 Polycrystalline structure on the ventral size of the *Cyanophrys remus* butterfly. (**a**) The butterfly *Cyanophrys acaste* is shown here in a resting position, when the hind wing overlaps the forewing. The ventral faces of the wings are displayed, with a cryptic dull green color. *Cyanophrys remus* displays identical resting position. (**b**) Ventral surface of the wing. One may remark the occurrence of bright spots colored blue, green, and yellow, which light up with different intensities as the illumination and observation conditions change. (**c**) Crosssectional transmission electron micrograph through an individual green scale; one may notice the precise regular structure within one grain and the different orientation of the neighboring grains

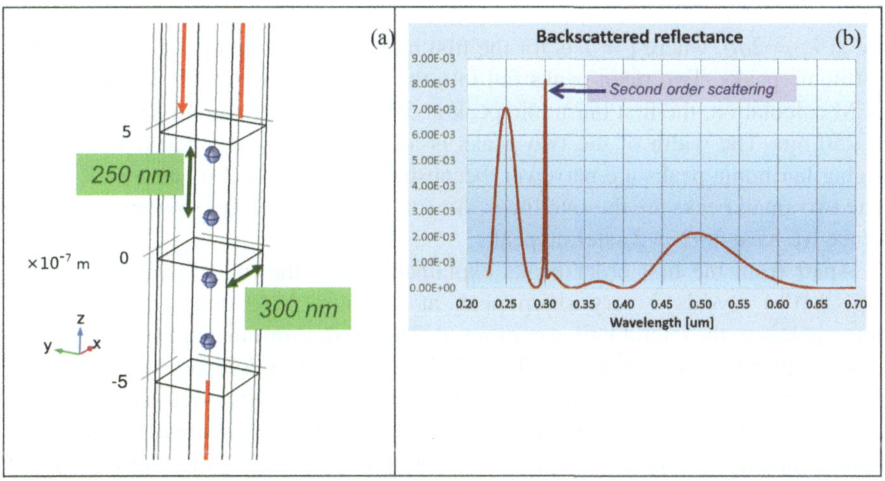

Fig. 8.10 Full 3D Maxwell equation calculation of the reflectance of a layered photonic nanostructure. (**a**) The unit cell of the calculation, containing four dielectric spheres of refractive index $n = 1.5$. The vertical distance of the spheres is 250 nm and the lateral size of the unit cell is 300 nm. The red arrows show the incoming, reflected, and transmitted wave vectors. (**b**) Calculated normal backscattered optical spectrum

Wave vectors of the incoming-, reflected-, and transmitted light are shown by red arrows. The incoming light arrives from the normal direction and the normal reflectance and transmittance was calculated as the function of the wavelength.

Figure 8.10b shows $R(\lambda)$, the normal reflectance as the function of wavelength. It has a wide peak around 500 nm, and another broadened, but sharper peak around 250 nm, two smaller intensity broad peaks between the two main peaks, and a sharp peak around 300 nm. The broad peaks can be reproduced by the first order scattering approximation, see Fig. 8.4 in Sect. 8.3. The main peaks of this spectrum

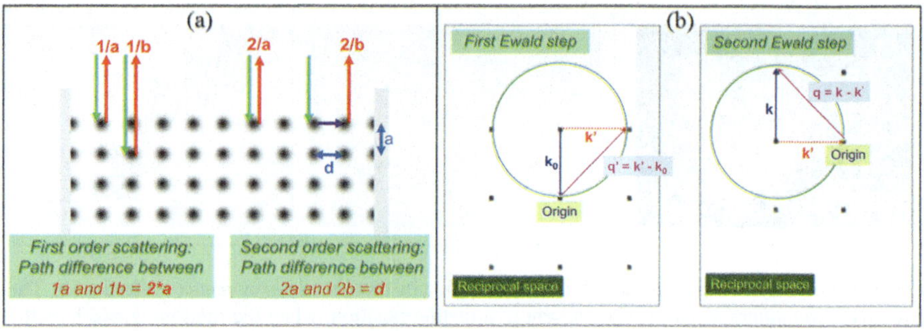

Fig. 8.11 First- and second order light scattering processes. (**a**) Real space lattice of four translation periodic layers with lateral periodicity d and layer distance a. The arrows show the light scattering processes, see the text for details. (**b**) Ewald sphere construction for the second order scattering in normal backstattered configuration. k_0, k, and k′ are the incoming, reflected, and intermediate wave vectors, respectively. See the text for details

are at $\lambda_j = 2d/j$, where j = 1 is for the first harmonic peak and j > 1 for the higher harmonic peaks. Two main peaks fall into the UV-VIS wavelength window of our FEM calculation, the first harmonic peak at 500 nm and the second harmonic peak at 250 nm. The width of the two peaks is identical in k scale, but in λ scale the higher harmonic peaks are narrower, because of the reciprocal relation of k and λ. The two small peaks are the satellite peaks (cf. Fig. 8.4). The layer number is n = 4, hence we have n−2 = 2 satellite peaks.

Apart from the first order peaks explained above, the full 3D calculation (cf. Fig. 8.10b) reveals a very sharp peak at 300 nm. Note that this wavelength corresponds to the lateral lattice constant d. This peak is originated from the second order scattering, as explained in Fig. 8.11. Figure 8.11a compares the first- and second order light scattering processes. In the first order normal backscattering the electromagnetic field components scattered on each layer are coherently added. Two such components, denoted with 1a and 1b are shown on Fig. 8.11a, those scattered from the first- and the second layer. As we can see on the figure, the path difference of the two light components is 2a, this defines the first order peaks. Light is scattered twice in second order processes. Such a process is shown by the vectors 2a and 2b. 2a is the component scattered by the first layer (same as 1a). Path 2b is a second order process. Light is first scattered on a sphere in the first layer and then on the adjacent sphere (at distance d) of the same layer, from where it is reflected back to the detector. The path difference of 2a and 2b is the lateral lattice constant, d, this gives rise to the peak at λ = d = 300 nm in Fig. 8.10b. This second order peak is sharp, because the structure is infinite in the lateral direction, contrary to the first order peaks, which are broad because of the finite vertical size of the system.

The two steps of the second order scattering are shown in Fig. 8.11b with the help of the Ewald construction. As was pointed out in [18], the probability of the second order scattering is the highest when the first order scattering direction is lateral, i.e. the scattered light remains inside the material for a long time, because it travels through the material parallel to the surface.

8.6 Conclusions

In this Chapter we invoked first order scattering theory in order to create a connection between the geometrical structure of butterfly wing scales and their optical spectrum. Optical spectrum of a butterfly scale is closely related to its reciprocal space structure. Indeed, butterfly scales are thin structures with moderate refractive index contrast, hence, the first Born approximation is applicable. Next we studied how order-disorder effects, which are always present in biological photonic nanostructures, influence their optical properties. A structure can deviate from a perfect long-range order in many different ways, i.e., there can be several kinds of randomnesses present in the structure. We analyzed in detail the effect of lateral- and vertical disorder, as well as polycrystalline disorder. The finite thickness of the butterfly scale causes a broadening of the spectral peaks and the appearance of small satellite peaks.

We also compared the first order scattering results with an exact electrodynamic calculation and identified the presence of second order processes, which cause the appearance of peaks in the UV part of the spectrum.

The simple reciprocal space approximation methods exposed in this chapter are useful because of their simplicity and tractable analytical properties. While exact calculations are readily available for periodic systems, the large unit cell necessary for modelling random structures has a high computational demand. First and second Born calculations can, however, easily handle large systems.

Acknowledgements This work was supported by the Hungarian NKFIH Grant Nos K 115724 and OTKA K 111741. G. I. M., and G. P. wish to thank the Hungarian Academy of Sciences and the Belgian FNRS for financial support.

References

1. Fox HM, Vevers G (1960) The nature of animal colours. Macmillan, New York
2. Berthier S (2007) Iridescences: the physical colors of insects. Springer, New York
3. Biró LP et al (2003) Role of photonic-crystal-type structures in the thermal regulation of a Lycaenid butterfly sister species pair. Phys Rev E 67:021907
4. Vukusic P, Sambles JR, Lawrence CR, Wootton RJ (1999) Quantified interference and diffraction in single Morpho butterfly scales. Proc R Soc Lond B Biol Sci 266:1403
5. Biró LP, Vigneron JP (2011) Photonic nanoarchitectures in butterflies and beetles: valuable sources for bioinspiration. Laser Photonics Rev 5:27
6. Kertész K, Molnár G, Vértesy Z, Koós AA, Horváth ZE, Márk GI, Tapasztó L, Bálint Z, Tamáska I, Deparis O et al (2008) Photonic band gap materials in butterfly scales: a possible source of "Blueparints". Mater Sci Eng B Solid-State Mater Adv Technol 149:259. ISSN 0921-5107
7. Vukusic P, Sambles JR, Ghiradella H (2000) Optical classification of microstructure in butterfly wing-scales. Photon Sci 6:61
8. Ghiradella H (1998) Microscopic anatomy of invertebrates, vol 11A (Locke M (ed)). Wiley-Liss, New York, pp 257–287

9. Yablonovitch E (1993) Photonic band-gap structures. J Opt Soc Am B Opt Phys 10:283
10. Jin C, Meng X, Cheng B, Li Z, Zhang D (2001) Photonic gap in amorphous photonic materials. Phys Rev B 63:195107
11. Vukusic P, Sambles JR (2003) Photonic structures in biology. Nature (London) 424:852
12. Kertész K, Bálint Z, Vértesy Z, Márk GI, Lousse V, Vigneron J-P, Rassart M, Biró LP (2006) Gleaming and dull surface textures from photonic-crystal-type nanostructures in the butterfly Cyanophrys remus. Phys Rev E 74:021922
13. Bálint Z, Kertész A, Moser A, Kertész K, Biró LP, Parker AR (2008) A supposition: structural colours resulting from both natural and sexual selection on an individual wing in the butterfly genusCyanophrys (Lepidoptera: Lycaenidae). Ann Hist-Natur Mus Nat Hung 101:63
14. Taflove A, Hagness S (2005) Computational electrodynamics: the finite-difference time-domain method. Artech House Publishers, Norwood
15. Benedek GB (1971) Theory of transparency of the eye. Appl Opt 10:459
16. Márk G, Vértesy Z, Kertész K, Bálint Z, Biró L (2009) Order-disorder effects in structure and color relation of photonic-crystal-type nanostructures in butterfly wing scales. Phys Rev E Stat Nonlinear Soft Matter Phys Rev E 80(5 Pt 1):051903
17. Prum RO, Quinn T, Torres RH (2006) Anatomically diverse butterfly scales all produce structural colours by coherent scattering. J Exp Biol 209:748
18. Maiwald L, Lang S, Jalas D, Renner H, Petrov AY, Eich M (2018) Ewald sphere construction for structural colors. Opt Express 26:11352–11365

Chapter 9
Prospects for Terahertz Imaging the Human Skin Cancer with the Help of Gold-Nanoparticles-Based Terahertz-to-Infrared Converter

A. V. Postnikov, K. A. Moldosanov, N. J. Kairyev, and V. M. Lelevkin

Abstract The design is suggested, and possible operation parameters are discussed, of an instrument to inspect a skin cancer tumour in the terahertz (THz) range, transferring the image into the infrared (IR) and making it visible with the help of standard IR camera. The central element of the device is the THz-to-IR converter, a Teflon® or silicon film matrix with embedded 8.5 nm diameter gold nanoparticles. The use of external THz source for irradiating the biological tissue sample is presumed. The converter's temporal characteristics enable its performance in a real-time scale. The details of design suited for the operation in transmission mode (in vitro) or on the human skin in reflection mode (in vivo) are specified.

9.1 Introduction

In searching for physical prerequisites to visualize a tumour, one can rely on the THz radiation's sensitivity to water, which is one of the most important components of the biological tissue. Hu and Nuss [1] were likely the first to point out the biomedical THz imaging, as they noted that the different water content of two different tissues (porcine muscle and fat) could yield a sufficient contrast. Even earlier, Ross et al. [2] and Chen et al. [3] have shown that the water content in cancerous tumours is higher than that in normal tissues. Further on, the water molecules absorb throughout the entire THz band (0.1–10 THz) [4]. These properties motivated a development of THz imaging the cancer cells in transmission geometry (for in vitro study of thin, clinically prepared tissue sample, see Fig. 9.1). In this approach, the necessary

A. V. Postnikov (✉)
Université de Lorraine, Metz, France
e-mail: andrei.postnikov@univ-lorraine.fr

K. A. Moldosanov · N. J. Kairyev · V. M. Lelevkin
Kyrgyz-Russian Slavic University, Bishkek, Kyrgyzstan
e-mail: altair1964@yandex.ru

A. Maffucci, S. A. Maksimenko (eds.), *Fundamental and Applied Nano-Electromagnetics II*, NATO Science for Peace and Security Series B: Physics and Biophysics, https://doi.org/10.1007/978-94-024-1687-9_9

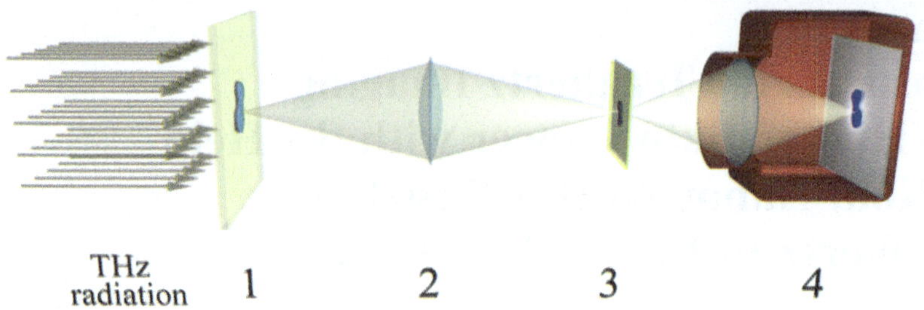

Fig. 9.1 Suggested setup for in vitro studies in the transmission mode. (1) a tissue sample; (2) THz objective (high-resistivity float zone silicon, high-density polyethylene or Teflon®) with a magnification M_1; (3) THz-to-IR-converter; (4) highly sensitive IR camera with magnification M_2. Malignant tissue eclipses the bright background

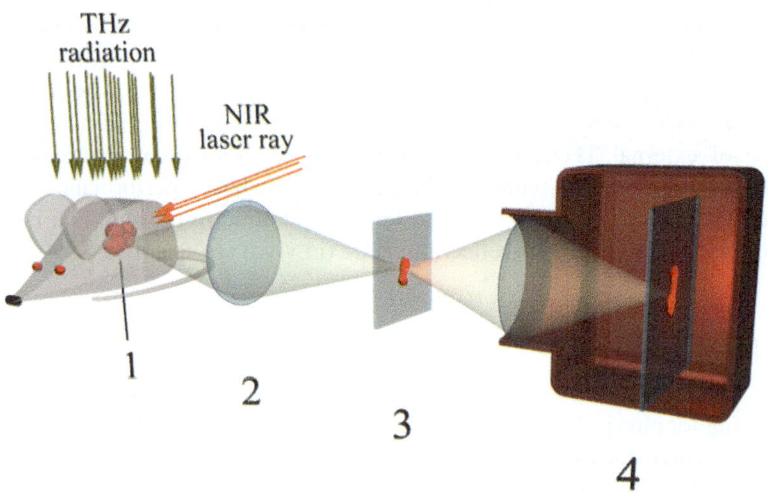

Fig. 9.2 Suggested setup for in vivo imaging in reflection mode. The near-infrared (NIR) laser serves for excitation of surface plasmons in GNPs inside a tumour (for heating water in cancer cells). (1) tumour; (2), (3), (4) the same as in Fig. 9.1. Malignant tissue gives a bright signal on a black background

contrast for imaging the tumour and discerning it from the normal tissue is provided by enhanced water content in cancer cells.

A fact that the water content of cancer cells is higher than that of normal cells, as well as the fact that the THz waves cannot penetrate moist tissue, favoured the development of another approach to the THz medical imaging, namely, the reflection geometry (for investigations in vivo, Fig. 9.2). Notably when imaging the human skin cancer, the reflection geometry is favoured over the transmission one, which, in its turn, allows to study the biological tissue samples of any organ's cancer.

The reflectance of THz radiation, and hence the contrast in imaging the area of cancer, is enhanced as the water temperature in cancer cells increases. This observation found its use in the THz imaging technique due to works of Rønne et al. [5] and Son [6], who studied the power absorption and the index of refraction of water depending on temperature. Their works prompted progress in optimizing the reflection geometry, e.g., by Oh et al. [7, 8], who demonstrated that the reflected THz signal grows with the water temperature in cancer cells. Further on, these findings helped to visualise tumour due to its larger contrast, compared with the background of normal cells in in vivo experiments [9–11].

To heat water in cancer cells, the gold nanoparticles (GNPs), like the targeted agents in photothermal therapy, are premediatedly delivered into the cancer cells, but not into the normal cells. The approach relies on the fact that the targeted agents, antibody-conjugated gold nanoshells [12] and solid gold nanospheres [13], are accumulated in cancerous tissue more efficiently than in the normal one. In photothermal therapy, these agents have been demonstrated to selectively kill cancer cells, leaving the normal cells unaffected [12–14]. Thereupon the tumour is non-invasively treated by irradiating with near-infrared (NIR) laser beam at ~650–1350 nm wavelength; this is the so-called "therapeutic window" where light has its maximum depth of penetration into the tissues. Under irradiation, the surface plasmons are excited in the GNPs; on dumping out the plasmons, the water is heated around the nanoparticles in cancer cells. In consequence, the cancer cells start to reflect the incident THz radiation even more efficiently, and thus can be more readily visualized by a highly sensitive IR camera (Fig. 9.2).

In works by Woodward et al. [15, 16], the feasibility of the THz imaging of the body with skin cancer in reflection geometry have been demonstrated.

Works by Woodward et al. [17] and Wallace et al. [18, 19] likely represent the pioneering studies that revealed a possibility of the THz imaging of the basal cell carcinoma. It was shown by Wallace et al. [19] and Pickwell and Wallace [20] that the maximum difference in refractive index between diseased and healthy tissue occurs at 0.35–0.55 THz. The fact that the maximum difference in absorption occurs in the vicinity of 0.5 THz strengthens the choice of the above frequency range as that providing the best imaging contrast when irradiating the biological tissue.

9.2 General Idea and Setup of the Terahertz-to-Infrared Converter

The availability of commercial IR cameras with temperature sensitivities of ~12–50 mK [21, 22] allows to suggest two schemes of THz imaging a tumour which are shown in Figs. 9.1 and 9.2. These imply an usage of the source of the THz radiation for irradiating the tissue sample in vitro or tumour in skin in vivo. In our work [23], we elaborated an idea that gold nanobars and nanorings irradiated by microwaves could become THz emitters with photon energies within the full width at half maximum (FWHM) of the longitudinal acoustic phononic density of states (DOS) of gold ($\approx$16–19 meV, i.e., 3.9–4.6 THz), with a maximum at $\approx$4.2 THz ($h\nu\approx$17.4 meV).

Further on, in Ref. [24] we have shown that gold nanorhombicuboctahedra could be used as emitters of radiation at “soft” (two-phonon difference frequency) 0.54 THz and at “hard” (summary frequency) 8.7 THz. The $\simeq 0.5$ THz emission is important for the THz imaging of human skin cancer, due to maximal contrast it yields between the cancer and the normal tissue, in the context of findings of Wallace and Pickwell [19, 20]. In the present work, the 0.38 THz radiation will be discussed, instead of 0.54 THz, in numerical estimates, for the reasons explained below. The 8.7 THz radiation, due to its shorter wavelength (see Table 9.1), might be advantageous for enhancing the spatial resolution of the image. The combined use of imaging at these two frequencies might be quite promising.

Such gold-nanoobjects-based THz sources are expected to be used in schemes of Figs. 9.1 and 9.2, therefore, below, when studying feasibility of the THz-to-IR gold-nanoparticles-based converter, we assume that the converter's nanoparticles absorb the THz photons of these three frequencies: 0.38, 4.2, and 8.7 THz. Some corresponding parameters are given in Table 9.1. The diffraction limit on the objective's resolution Δx can be estimated as $\Delta x \approx \lambda \cdot l / A$ (λ: wavelength, l: distance between the objective centre and the object, hence patient's skin; A: the objective's aperture). The estimates show that these frequencies could yield acceptable resolutions for studying the structure of, say, pigmentary skin nevi. Further on, in Table 9.1, the transmissions of the 0.1 mm thick Teflon® film at frequencies of interest are given. Below, it will be shown that this material could be a good choice for the THz-to-IR-converter matrix, in accordance with data of Ref. [25].

In our work [26], a physical mechanism has been suggested for heating the GNPs with radiofrequency radiation. At the core of this mechanism are longitudinal acoustic vibration modes (LAVMs), which a priori could have played a role also in the absorption of THz photons. In the Sect. 9.3.1 below, we consider the special case of absorption of THz photons by GNPs without involvement of LAVMs, so that the momentum conservation law is fulfilled due to uncertainty in momentum of the Fermi electrons of gold.

First, we consider the transmission geometry setup (Fig. 9.1). The idea implies that an image (in the THz rays) of the tissue sample (position 1 in Fig. 9.1) is projected by the objective (position 2 in Fig. 9.1) onto the two-dimensional THz-to-

Table 9.1 Operating frequencies of the THz-to-IR converter and corresponding objective's resolutions

Frequency (THz)	Wavelength (μm)	Photon energy (meV)	Photon momentum (10^{-25} g·cm·s^{-1})	Objective's resolution (μm) at $l\sim$100 mm	Transmission k_1 of the 0.1 mm thick Teflon® film (%)
0.38	789.5	1.57	0.84	~395.0[a]	~95
4.2	71.3	17.4	9.3	~71.3[b]	~85
8.7	34.5	36.0	19.2	~34.5[b]	~70

[a] $A \sim 200$ mm
[b] $A \sim 100$ mm

IR-converter, which is a matrix, transparent both in THz and IR rays, with embedded GNPs (position 3 in Fig. 9.1). The GNPs, on irradiation with THz rays, convert the energies of THz photons into heat, being so the bright spots for subsequent detection by the IR-camera (position 4 in Fig. 9.1).

The idea of cancer cells detection is that the latter, rich in water, will strongly absorb the THz radiation; consequently, the corresponding (eclipsed) areas in the projected image on the detector plane will generate less heat and appear dark in the resulting IR image, over the otherwise bright background.

Now we turn to the reflection geometry (Fig. 9.2). In this setup, it is additionally assumed that the GNPs are delivered to the tumour in advance like the targeted agents in photothermal therapy, and are irradiated in situ with the NIR laser. The laser's radiation penetrates the skin and the tissue and excites surface plasmons in the therein implanted GNPs. The surface plasmons heat the tumour, that in its turn enhances reflection of the THz radiation (that is known to be efficiently scattered by water at elevated temperatures) from it. The THz rays reflected from the tumour are focused by a lens on the THz-to-IR converter's matrix, creating in this plane a (bright) THz image of the tumour. Further on, like in the previous case depicted in Fig. 9.1, the GNPs within the THz-to-IR converter absorb the THz radiation, generate heat and become IR sources, producing a bright image detectable by the IR camera.

Concerning the realisation of the THz-to-IR converter, two schemes shown in Fig. 9.3 may come into discussion. That in the form of a thick film with embedded GNPs (Fig. 9.3, lower scheme) seems preferable over single-layer deposition (Fig. 9.3, upper scheme), because it allows to achieve larger "projected" density of GNPs per surface unit, avoiding at the same time to place them too closely. We'll see that these both considerations are important. The spatial resolution and the depth of focus of the IR camera are the essential parameters to guide the optimal design.

The following considerations help to specify the lens system and sizes of the THz-to-IR converter:

(1) The objective of the IR camera ensures the close up operation mode with a magnification $M_2 = 1$. For infrared cameras, which are kept in mind [21, 22], this imposes a short distance (~22–23 mm) between the "object" (i.e., the THz-to-IR converter's matrix with GNPs) and the objective's edge, as well as small

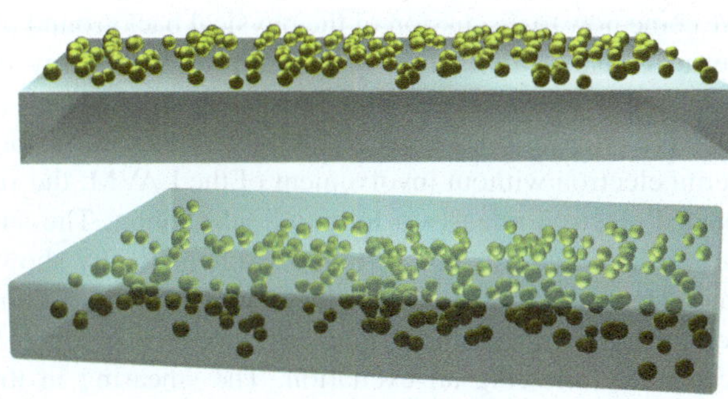

Fig. 9.3 Schemes of the THz-to-IR converter, in the form of a substrate transparent in THz with GNPs deposited onto it (above), or in the form of a matrix transparent in THz wavelength range with embedded GNPs (below)

sizes (9.6 mm × 7.7 mm) of the "object". The THz-to-IR converter being so small seems technologically advantageous. At small sizes of the converter's matrix, the distortions caused by the first lens (the THz objective) would be small, too. The second lens (the IR camera objective) is simplistically shown in Figs. 9.1 and 9.2, whereas its real structure may be much more complex.

(2) It is convenient to assume the THz objective to be changeable, with different magnifications, say, $M_1 = 1$; 0.2; 5, whereby the two latter values can be obtained by inverting the same objective. Magnification of $M_1 = 0.2$ would serve a preliminary examination of the patient's skin area, over the field of view (FOV) of about (9.6 mm × 7.7 mm)/0.2 = 48 mm × 38.5 mm. Operating with the magnification of $M_1 = 1$, one could inspect a chosen object within the FOV of 9.6 mm × 7.7 mm (for example, a pigmentary skin nevi). The magnification $M_1 = 5$ with its corresponding FOV of (9.6 mm × 7.7 mm)/5 = 1.92 mm × 1.54 mm would enable inspecting details of the pigmentary skin nevi.

The diffraction limits on the objective's resolution Δx are given in Table 9.1. At frequency 4.2 THz and $M_2 = 1$ (in the *close up* operation mode) and pixel size $d = 15\,\mu\text{m}$ this means that the linear "uncertainty" of the image is covered, on the average, by (~71.3 μm)/(~15 μm) ≈ 5 pixels of the IR camera detector, that seems acceptable to resolve a meaningfully detailed pattern. At frequencies 0.38 and 8.7 THz and $M_2 = 1$ the linear "uncertainties" would be covered, on the average, by ≈26 and 3 pixels, respectively.

The lenses with desirable parameters (focal distance, diameter) for the THz range, made out of various materials, can be selected from the lists of commercially available products – see, e.g., Ref. [27].

9.3 Energy and Momentum Conservation Relations Within the Photon – Electron – Phonon System

9.3.1 Absorption of THz Photons by GNPs Helped by Uncertainty of the Fermi Electrons' Momenta

We come now to discussion of the physical background of the THz to IR conversion. Let us consider a situation in GNP, when the energy interval between the Fermi level and a nearby accessible vacant electron energy level equals the energy of the THz photon, $h\nu$. Figure 9.4 depicts a case of absorption of a THz photon by Fermi electron without involvement of the LAVM; the subsequent relaxation of the excited electron releases a longitudinal phonon. The surface of revolution around the vertical (energy) axis in Fig. 9.4 schematically shows the energy dispersion of free electrons (as function of just two momentum coordinates), delimited at the bottom by Fermi momentum/energy and at the top – by a discrete energy level accessible following an excitation. The smearing in the upper momentum plane

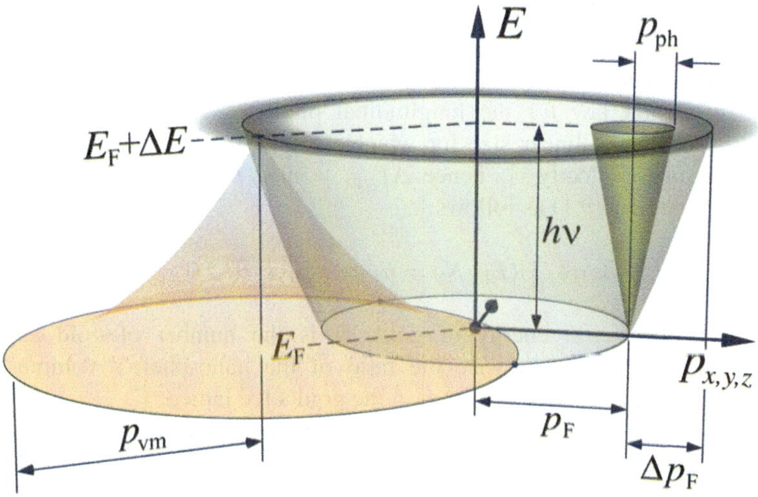

Fig. 9.4 Scheme of absorption of THz photon by the Fermi electron (right part of the figure) and the latter's subsequent relaxation via releasing a phonon (left part of the figure). The dispersion law of an electron is depicted by ascending paraboloid, that of the photon – by narrow ascending cone, that of longitudinal phonon – by descending bell mouth. Dispersions are shown over two-dimensional momentum plane. The crossing points of the electron and phonon dispersion surfaces at $E = E_F$, marked by dots, indicate relaxation points for the excited electron. See text for details

indicates the uncertainty of the electron momentum due to spacial confinement (see below), that helps to match the momentum conservation condition involving a photon. The dispersion of a THz photon is shown by a narrow cone, starting from some momentum/energy value of a Fermi electron. In the follow-up of an excitation to the energy $E_F + \Delta E$, the electron can relax releasing a longitudinal phonon. The phonon dispersion is depicted by a descending (inverted) bell mouth, starting from some momentum/energy of the excited electron. Essential observations are that (i) for small enough nanoparticles, the momentum and energy conservation in the course of an immediate electron excitation by a THz photon can be assured by the uncertainty relation; (ii) following the relaxation of an excited electron via releasing a phonon, the electron's momentum may undergo a large reorientation (largely retaining its magnitude). Let us now consider some details more attentively.

First we try to estimate the sizes of gold nanospheres which could be efficiently used in the THz-to-IR converter. The nanoparticle diameter D allows to "tune" the electron energy separation ΔE_{el}, according to the Kubo formula for the level spacing in a nanoparticle, Kubo [28, 29] (see also Appendix 1 in Postnikov and Moldosanov [26]). The energy delivered by a THz photon $h\nu$ should match an integer number of steps, m_{el}, in the electron excitation ladder; simultaneously it should fit an integer number of energy steps n_{vm} for the vibration mode, ΔE_{vm}:

$$m_{\text{el}}\,\Delta E_{\text{el}} = h\nu = n_{\text{vm}}\,\Delta E_{\text{vm}}\,. \tag{9.1}$$

For simplicity and order-of-magnitude estimates, we assume for the beginning a linear dispersion law for the longitudinal phonons, supposing that they are propagating along the linear size for which we take the diameter D, possessing the nominal sound velocity υ_{L}, hence $\Delta E_{\text{vm}} = \upsilon_{\text{L}}h/D$. A reference to the Kubo formula modifies Eq. (9.1) as follows:

$$m_{\text{el}}(4/3)(E_{\text{F}}/N) = h\nu = n_{\text{vm}}\cdot\,\upsilon_{\text{L}}\cdot(h/D)\,, \tag{9.2}$$

where E_{F} is the Fermi energy of gold, N is the number of gold atoms in the nanosphere, that is otherwise the ratio of the nanosphere's volume $V = (4/3)\pi(D/2)^3$ to the volume per atom in the gold's fcc lattice, $V_{\text{at}} = a^3_{\text{Au}}/4$, hence $N = V/V_{\text{at}} \approx 2.09\cdot(D/a_{\text{Au}})^3$. With this, the diameter D is expressed via the m_{el} and n_{vm} parameters as follows:

$$D \approx 0.798\,a^{3/2}_{\text{Au}}\left(\frac{m_{\text{el}}}{n_{\text{vm}}}\,\frac{E_{\text{F}}}{\upsilon_{\text{L}}h}\right)^{1/2}, \tag{9.3}$$

or, with the material constants inserted and specifying the units,

$$D \approx 4.23\cdot(m_{\text{el}}/n_{\text{vm}})^{1/2}\ \text{nm}. \tag{9.4}$$

It follows from the right parts of Eqs. (9.1) and (9.2) that

$$\nu = (n_{\text{vm}}\Delta E_{\text{vm}})/h = n_{\text{vm}}(\upsilon_{\text{L}}/D)\,. \tag{9.5}$$

For given D and ν, we can hope that, by force of Eqs. (9.4) and (9.5), certain combinations of m_{el} and n_{vm} would emerge as "resonance" ones. The optimal choice of all related values comes about from the following considerations:

(i) For securing a sufficient contrast of the image of cancer/normal tissue, the frequency ν ought to be in the range 0.35–0.55 THz.

(ii) The diameter D must exceed $\simeq$ 8 nm, in order to prevent the GNP's cooling via spontaneous emission of THz photons. This observation was guided by our study [30] dedicated to a suggested explanation of the size effect in the heterogeneous catalysis on GNPs. Nanoparticles with sizes inferior to 7 nm tend to cool down due to a spontaneous emission of THz photons, hence become useless if the objective is opposite, to *heat* the GNPs in the THz-to-IR converter by incoming THz radiation. On the other side, an increased D would degrade the converter's sensitivity, since an elevated THz power level will be required to efficiently heat the GNPs.

(iii) m_{el} and n_{vm} must be integer. With respect to the latter condition, we acknowledge that it can be imposed only approximately, because the exact quantisation criteria would be difficult to control under practical variations of GNPs' shapes

Table 9.2 Parameters of gold nanospheres with thermal conductivity $\lambda_{1p} = 73.65\ \mathrm{W{\cdot}m^{-1}{\cdot}K^{-1}}$ and diameter $D = 8.5$ nm suitable for detection of soft and hard THz radiation. Estimations are done for $m_{\mathrm{el}} = 4$, $n_{\mathrm{vm}} = 1$. See text for details

Frequency (THz)	Phonon momentum (h/a_{Au})	$(10^{-20}\ \mathrm{g{\cdot}cm{\cdot}s^{-1}})$	Skin depth (nm)	Δp_{D} $(10^{-21}\ \mathrm{g{\cdot}cm{\cdot}s^{-1}})$	Δp_{F} $(10^{-23}\ \mathrm{g{\cdot}cm{\cdot}s^{-1}})$
0.38	0.048	0.78	121.05	≥ 1.24	1.79
4.35	0.770	12.5	35.8		41.1

and sizes. Still, the smaller the $(m_{\mathrm{el}}, n_{\mathrm{vm}})$ the stronger the importance of the confinement effects and hence of the resonance conditions in the energy matching. We note moreover that the estimation of ΔE_{vm} in terms of v_{L} is only approximate. Anyway, accepting the nominal value of the velocity of sound in gold $v_{\mathrm{L}} = 3.23{\cdot}10^5$ cm/s, we arrive at the following compromise concerning the parameter values: $\nu = 0.38$ THz, $D \approx 8.5$ nm; $m_{\mathrm{el}} = 4$; $n_{\mathrm{vm}} = 1$, that was accepted for the following analysis.

Last but not least, the skin depth in gold at frequencies of interest exceeds this GNP size considerably (see Table 9.2). One can therefore assume that the electric field penetrates the volume and is of the same strength throughout the nanoparticle. Moreover, the GNP size should be smaller than the mean free path of electrons in gold, which is, at 300 K, $l_{\mathrm{b}} \approx 13.1$ nm [31], in order to facilitate the estimation of the thermal conductivity.

In addition to the above arguments based on the energy conservation law, we should pay attention to issues of the momentum conservation. On excitation of a Fermi electron by a photon of energy $h\nu$, the electron momentum (assuming a free-electron dispersion law) gets an increase. The momentum of the absorbed THz photon, p_{ph}, is much smaller and hence cannot help to overcome the mismatch. However, at small enough GNP sizes, the uncertainty in the electron's momentum might well absorb the value of Δp_{F}.

Namely, the modification (an increase) of the Fermi electron's momentum in the course of absorbing a THz photon is $\Delta p_{\mathrm{F}} \approx h\nu/v_{\mathrm{F}}$, where $v_{\mathrm{F}} \approx 1.4{\cdot}10^8$ cm/s is the Fermi velocity of electrons in gold [32], hence $\Delta p_{\mathrm{F}} \approx 4.73{\cdot}10^{-35}{\cdot}\nu$ g·cm. This should be compared to the Heisenberg uncertainty of the electron's momentum in a confined geometry,

$$\Delta p_D \geq h/(2\pi D)\,,$$

whence, with Eq. (9.4) and the parameter values for gold, we get:

$$\Delta p_D \geq 2.49{\cdot}10^{-21}(n_{\mathrm{vm}}/m_{\mathrm{el}})^{1/2}\ \mathrm{g \cdot cm \cdot s^{-1}}\,.$$

Some numerical estimations are given in Table 9.2. It follows that, for the GNP size we consider, the condition $\Delta p_{\mathrm{F}} \leq \Delta p_D$ is by far respected. Therefore, the mismatch of the electron momenum to satisfy the momentum conservation on absorbing a

THz quantum with frequency from the range of interest for biomedical applications would be always "absorbed" by the Heisenberg uncertainty relation.

9.3.2 GNP Heating Through Release of Longitudinal Phonons

The relaxation of excited electron via releasing longitudinal phonon(s), that would result in heating the GNP, is illustrated by a "descending" scheme in the left part of Fig. 9.4. From the **energy conservation** condition, the excitation energy $\Delta E = m_{\text{el}} \cdot \Delta E_{\text{el}}$ should match the energy of a vibration mode, according to Eq. (9.1). Figure 9.4 depicts for simplicity a single-phonon process. One can moreover imagine two-phonon (or, in principle, multiphonon) processes, following the electron excitation by a "hard" ($\sim$8.7 THz) photon.

From the side of **momentum conservation**, it is seen from Fig. 9.4, in which at least the qualitative relations are roughly respected, that there is no general problem of momentum mismatch like that existing in the case of absorption of a photon, when the photon dispersion surface (a narrow cone) was entirely placed inside the free-electron paraboloid. Indeed, in case of releasing a phonon, the *minimal* momentum mismatch, i.e., the opening of the bell mouth on the left of Fig. 9.4 for the case $n_{\text{vm}} = 1$, is $p_{\text{vm}} \simeq h/D$, hence $\simeq 2\pi$ times larger than Δp_D listed in Table 9.2 (that makes $7.8 \cdot 10^{-21}$ g·cm·s^{-1}, for $D = 8.5$ nm), and by far larger than Δp_{F}. The momenta values of "genuine" longitudinal phonons having frequencies of interest, as determined from experimental phonon dispersion along the ΓX direction [33] and its numerical fit given in Ref. [24], are also shown in Table 9.2. As could be expected, for $\nu = 0.38$ THz the "exact" (experimental) value perfectly fits that from the simple linear dispersion model as estimated above. Therefore the matching conditions for both energy and momentum, represented by intersections of respective dispersion-law surfaces, can typically be easily found. Figure 9.4 shows two such "hits", indicated by thick dots, on the circle of radius p_{F} in the basal plane $E = E_{\text{F}}$. In 3-dimensional reciprocal space, such matches will be placed on continuous lines of intersection of isoenergetic surfaces describing the electron and phonon dispersions. Quantum confinement conditions for electrons and phonons will select distinct "spots" along such lines, which will be however smeared by force of the uncertainty relation and the particles' irregularity and dispersion of sizes. As already mentioned above in relation with the energy conservation, it is difficult to elaborate on specific relations in the most general case; however it seems plausible that the electron/phonon energy/momentum matches can be, in principle, easily found and will "work" in the relaxation mechanism.

A quite general and not trivial observation concerning the energy/momentum conservation in the course of electron-phonon relaxation is that the momentum of the relaxing electron may be scattered, in the process, quite far along the Fermi surface. This follows from the comparison of typical values of Δp_{F} and (its superior by orders of magnitude) phonon momentum in Table 9.2.

Summarizing, the heating of GNP may occur thanks to a combination of two circumstances. (i) The absorption of the THz photon is accompanied by an excitation of a Fermi electron, which is only made possible by uncertainty of the latter's momentum. (ii) The relaxation of the excited electron brings about release of a phonon (or, several phonons), in which process the uncertainty in the momentum of the longitudinal phonon (or, the combined uncertainty in case of several phonons) may also play an auxiliary role as it mitigates the exact energy/momentum matching conditions, however, this is not qualitatively essential. The channeling of initial photon energy into a vibration channel amounts to heating the particle.

9.4 Estimations of Parameters of the THz-to-IR Converter

For further quantitative estimates, we consider a model of matrix-based converter (Fig. 9.3, lower scheme), with Teflon® and silicon as possible matrix materials, transparent in both THz and IR wavelength ranges (specifically, throughout ~1–5 μm in IR), and gold nanospheres embedded therein.

As was mentioned above, according to the data of Ref. [25], the good transmission of Teflon® film of 0.1 mm thickness at the THz wavelengths of interest makes this material promising as a substrate to host GNPs. The transmission within the IR spectral range (about 80–90% for IR wavelength of 3–5 μm and about 50% for 9 to 12 μm) is also acceptable. For the infrared camera, the Mirage 640 P-Series [21] (specifically, the version operating in the wavelength range from 3 to 5 μm) could be a good choice, the FLIR A6700sc MWIR [22], operating in the same wavelength range, being a reasonable alternative.

Using the previously identified optimal GNP diameter, we calculate now the temporal characteristics and spatial distribution of heat generated within an isolated GNP embedded in the matrix, as well as the power levels of the THz radiation required to maintain the GNPs in thermal equilibrium at, or above, the temperature sensitivity threshold of IR camera.

9.4.1 Radiation Power Required to Hold GNPs at Temperatures Defined by Their Emissivity Factor α

The temperature distribution over, and around, the GNP can be described by the heat conduction equation in spherical symmetry with the source function $q(r)$ taken into account:

$$\rho C \frac{\partial T}{\partial t} = \frac{1}{r^2} \frac{\partial}{\partial r} \left(\lambda r^2 \frac{\partial T}{\partial r} \right) + q(r)\,, \tag{9.6}$$

where $T(t, r)$ is the temperature, ρ the volume density, C the specific heat, λ the thermal conductivity, q the volume density of the heat source, $q = Q/[(4/3)\pi R_0^3]$ (Q being the power at the heat source); T depends on the radius r and time t.

We assume the thermal parameters to be temperature independent and uniform inside and outside the particle, as follows:

$$\begin{array}{lllll} 0 \le r \le R_0: & \lambda = \lambda_1, & \rho = \rho_1, & C = C_1; & q = Q/[(4/3)\pi R_0^3], \\ R_0 \le r \le R: & \lambda = \lambda_2, & \rho = \rho_2, & C = C_2; & q = 0, \end{array} \tag{9.7}$$

where R_0 is the nanoparticle's radius ($R_0 = 4.25$ nm, as argued in Sect. 9.3.1) and R is the radius of the Teflon® shell, $R \gg R_0$. For practical calculations below, $R = 500$ nm was taken as an effective "infinity", in view of much larger factual matrix thickness (~0.1 mm) and the assumption that individual GNPs are too distant to interfere. The initial and edge conditions were as follows:

$$T(0, r) = T_R; \qquad \left.\frac{\partial T(t, r)}{\partial r}\right|_{r=0} = 0; \qquad T(t, R) = T_R. \tag{9.8}$$

A solution for Eqs. (9.6)–(9.8) was sought for by the numerical method of lines relative to $\Delta T = T - T_R$, where $T_R = 300$ K, as was described in detail in Ref. [34].

For nanoparticles, both their specific heats and thermal conductivities depend on the nanoparticle's size (see Table 9.3). According to Gafner et al. [35], the specific heat of GNPs is just slightly above that of the bulk gold. We extrapolated the specific heat C_{1p} of the 8.5 nm diameter GNP from the data of Ref. [35] and estimated it to be 133.7 J·kg^{-1}·K^{-1}. The difference in the thermal conductivity is much more dramatic. In fact, the *electron* thermal conductivity on the nanoscale is much higher than the *lattice* thermal conductivity [36, 37], however lower than that of the bulk metal. When the GNP's characteristic size (diameter) D is smaller than the mean free path of electrons in the bulk gold, i.e., $l_b \approx 36.1$ nm at 300 K [31], the thermal conductivity of the nanoparticle λ_{1p} can be estimated as follows [31]: $\lambda_{1p} \approx (\lambda_{1b}/l_b)\cdot D \approx 73.65$ Wm^{-1}K^{-1}, where $\lambda_{1b} = 312.8$ Wm^{-1}K^{-1} is the thermal conductivity of bulk gold.

The source term in Eqs. (9.6) and (9.7), stationary in our assumption, accounts for the power Q delivered to the particle. Obviously dependent on the power of the primary THz source, the relevant (threshold) values of Q for our analysis are

Table 9.3 Parameters of materials used in numerical solutions of the heat equation

Material	Volume density (g·cm^{-3})	Specific heat (J·kg^{-1}·K^{-1})	Thermal conductivity (W·m^{-1}·K^{-1})
Nanoparticle : $D = 8.5$ nm gold sphere	$\rho_1 = 19.32$	$C_{1p} = 133.7$	$\lambda_{1p} = 73.65$
Matrix : Teflon®	$\rho_2 = 2.12$	$C_2 = 1160$	$\lambda_2 = 0.26$
Matrix : Silicon	$\rho_2 = 2.33$	$C_2 = 712$	$\lambda_2 = 159$

Table 9.4 Parameters of THz-to-IR converter based on GNPs with the radius $R_0 = 4.25$ nm, for two values of emissivity factor α

		Teflon matrix		Silicon matrix	
α	ΔT_α (mK)	Q_T (nW)	n_T	Q_Si (nW)	$n_\mathrm{Si}(\times 10^{-3})$
1	12	0.168	0.23	49.2	0.79
0.5	24	0.335	0.23	98.4	0.79

those which enable a heating, as a steady solution of Eq. (9.6) – i.e., after an infinite saturation time, – of the (embedded) GNPs to temperature levels detectable by the IR camera. The dynamics of the heating will be discussed below; for the reasons elaborated in the next section, we fix such reference threshold value at 12 mK, that is the reported temperature sensitivity of the *Mirage 640 P-Series IR* thermal imaging camera [21]. Individual GNPs, however, may largely vary in their ability to convert absorbed power into heat (depending on the particle's roughness etc.), that can be grasped into a phenomenological emissivity factor α, varying between 1.0 (ideal Q to ΔT conversion) and almost zero. Without a clue as for the actual values of α, we consider in the following, for reference purposes, the values $\alpha = 1$ and $\alpha = 0.5$, the generalisation being straightforward. The threshold ΔT_α values needed to yield visibility by the IR camera will be scaled by $\times \frac{1}{\alpha}$, hence 12 and 24 mK, correspondingly, for the α's under discussion. The estimated powers Q_T and Q_Si, defined by GNPs' emissivity factors α, versus the temperature rise ΔT_m for GNP of radius $R_0 = 4.25$ nm in Teflon® and silicon spherical shells are given in Table 9.4. In Fig. 9.5, the radial distributions of temperature around GNPs embedded in the matrix are shown.

9.4.2 Estimations of THz Power Threshold Sensitivity for the THz-to-IR Converter + IR Camera Operation

On having discussed the "performance" of a single embedded GNP for the THz-to-IR conversion, we turn to an assessment of a realistic device, to be composed of distributed GNPs and "viewed" by practically available IR cameras. Let us consider a spot on the converter plate, which has to be mapped onto the pixel in the IR camera's focal plane array (FPA). As was mentioned in the beginning of Sect. 9.2, the IR camera operates in the close up mode, so that its objective's magnification equals nearly 1. Assuming that d is the pixel size, the spot area is $\sim d^2$. The Stefan – Boltzmann law yields the energy flux from the area radiating as the absolutely black body (i.e., the radiant flux surface density) into a semisphere (solid angle of 2π), in the wavelength range from 0 to ∞, as $\varepsilon_{bb} = \sigma T^4$, where $\sigma = 5.67 \cdot 10^{-8}\,\mathrm{Wm^{-2}K^{-4}}$ is the Stefan – Boltzmann constant, and T the surface temperature (taken $T = 300$ K in the following). Let us assume that ΔT_{bb} is the nominal temperature sensitivity of the IR camera. This means that the detectable excess (over the background) in the

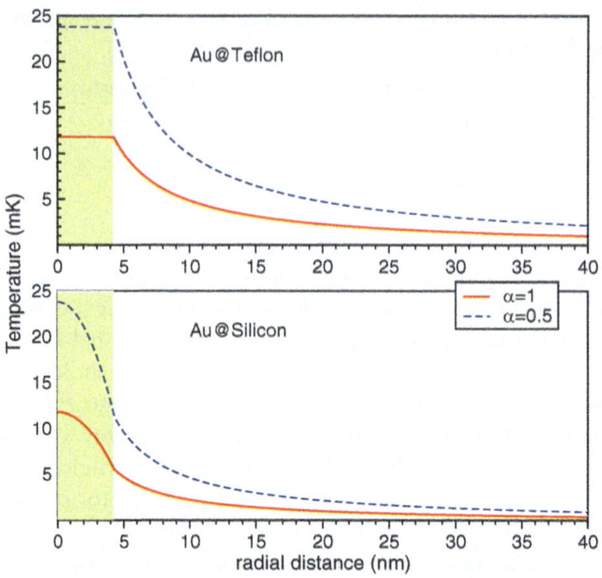

Fig. 9.5 Radial distributions of excess temperatures throughout the GNP of radius $R_0 = 4.25$ nm (marked by a colour bar) and its embedding by Teflon® and silicon spherical shells, for the values of Q_{T} and Q_{Si} from Table 9.4, corresponding to emissivity factor $\alpha = 1$ and $\alpha = 0.5$

surface density of the radiant flux emitted by the reference spot into the solid angle 2π within the whole wavelength range needs to be at least

$$\Delta\varepsilon_{bb} = 4\sigma\, T^3 \Delta T_{bb}\,, \tag{9.9}$$

in order to cause a response of the IR camera. We assume further on that the surface spot is expanded into a volume element with the same area d^2 and thickness δ, inferior to the depth of focus l_{dof} of the IR camera's objective. We specify that the volume element contains n GNPs with the emissivity factor α, heated by absorbing the THz energy. Those GNPs within $\delta \leq l_{\mathrm{dof}}$ heated up to a temperature of 300 K+$(\Delta T_{bb}/\alpha)$ will be perceived by the IR camera.

The properties of the substrate, in what regards the heat exchange, need to be grasped by the model to enable realistic estimates of the temporal characteristics and spatial resolution. Numerical estimations have been done as explained above, assuming that the GNP is placed into a spherical shell of Teflon® (or silicon) of 500 nm radius, whereby the power of the THz radiation Q heats the GNP in a steady-state mode up to the temperature of $(\Delta T_{bb}/\alpha)$ above the background – see a solution of the heat conduction equation (9.6) in Sect. 9.4.1, with the Q values for the source term, Eq. (9.7), as given by Table 9.4. In the steady-state heat transport, the surface density of the radiant flux generated by n GNPs element and emitted outwards within the solid angle 4π and the wavelength range from 0 to ∞ is $(k_2 \cdot n \cdot Q)/d^2$, where k_2 is the transmission of the Teflon® matrix within the operating wavelength range of the IR camera. Half of this, i.e., $(k_2 \cdot n \cdot Q)/(2d^2)$, will be emitted in the direction towards the IR camera. Comparing this latter value against the threshold radiant flux from Eq. (9.9) specifies the minimal number of GNPs needed in the volume element so that the resulting radiant flux surface density be sufficient to cause the IR camera's response:

$$n = \frac{8\sigma T^3 \Delta T_{bb}\, d^2}{k_2 Q}, \quad \text{or more generally} \quad n = \frac{8\sigma T^3 \Delta T_{\alpha}\, d^2}{k_2 Q}, \tag{9.10}$$

where $\Delta T_{\alpha} = \Delta T_{bb}/\alpha$ (specified in Table 9.4) accounts for the emissivity factor α being different from 1.

The Q values to use in Eq. (9.10), as they follow from calculations described in Sect. 9.4.1, are listed in Table 9.4 for two trial levels of the emissivity factor α of GNPs. The Q_{T} and Q_{Si} are the steady-state THz powers to be delivered to a GNP placed inside the spherical shell made of, respectively, Teflon® or silicon, in order to heat it from temperature $T_0 = 300\,\mathrm{K}$ to $T_0 + \Delta T_{bb}/\alpha$. To be specific, the value ΔT_{bb} was chosen to be 12 mK, the reported temperature sensitivity of the *Mirage 640 P-Series* infrared thermal imaging camera [21].

In the model considered, the minimal possible thickness of the film matrix is 1 μm, the diameter of the spherical shell covering the GNP. For further assessments, Teflon® film of 0.1 mm thickness would be an acceptable choice because this is less than the typical depth of focus l_{dof} (~0.3 mm). Moreover, the transmission data k_1 and k_2 in, correspondingly, THz and IR ranges are available for the Teflon® film of namely 0.1 mm thickness [33] – see the Table 9.1, where the k_1 values are given for three frequencies of interest, whereas $k_2 \approx 85\%$ within the wavelength range from 3 to 5 μm, that is, the operating range of the *Mirage 640 P-Series* infrared camera.

As follows from Table 9.4, the values of n_{T} or n_{Si}, i.e., the "threshold" numbers of GNPs within the volume element that maps onto a pixel of the IR camera, are by far less than 1 in all combinations of the parameters considered. In order to "sensibilize" each pixel, the number of GNPs within the corresponding volume element must obviously be, at least, one. An increase in the GNPs concentration beyond this number ensures each pixel to be successfully addressable.

Assuming n_{T} or $n_{\mathrm{Si}} = 1$, the concentration of GNPs in the substrate must be $N_{\mathrm{T}}^* = N_{\mathrm{Si}}^* = 1/(d^2 \cdot \delta)$. Taking the *Mirage 640 P-Series* IR camera detector as an example, the pixel size is $d = 15\,\mu\mathrm{m}$, therefore, for the above justified thickness of the Teflon® layer being 0.1 mm, the GNP concentration is $\approx 4.44 \cdot 10^4\,\mathrm{mm}^{-3}$. This will suffice to make every pixel addressable.

We can further on estimate the power of THz radiation source needed for a Teflon® matrix of the 9.6 mm×7.7 mm size and 0.1 mm thickness to work. The total amount of the GNPs in the matrix is $N_T^* \cdot 9.6\,\mathrm{mm} \times 7.7\,\mathrm{mm} \times 0.1\,\mathrm{mm} \approx 4.44 \cdot 10^4\,\mathrm{mm}^{-3} \cdot 7.39\,\mathrm{mm}^3 \approx 3.28 \cdot 10^5$. Typically for rough estimates one takes the emissivity factor $\alpha = 0.5$, then the power needed to heat such GNP by 24 mK is 0.335 nW, and the required power of the whole THz source would be $3.28 \cdot 10^5 \cdot 0.335$ nW $\approx$ 110 μW. For silicon matrix, the corresponding values of power are 98.4 nW and $3.28 \cdot 10^5 \cdot 98.4$ nW $\approx$ 32.3 mW, respectively, i.e., larger than those for the Teflon® matrix by a factor of ≈300.

The values 110 μW and 32.3 mW can be considered as the conventional powers required for operating the THz-to-IR converters built around the Teflon® and silicon matrices, respectively, in the transmission mode (Fig. 9.1). For the reflection mode (Fig. 9.2), these values could classify a sensitivity of the "THz-to-IR converter + IR camera" system.

9.4.3 Temporal Characteristics of the THz-to-IR Converter + IR Camera System

Being also interested in the detector's reaction time, we turn now to time-dependent solution of Eq. (9.6), using the "threshold" Q values from Table 9.4. The time evolutions of temperature at the center of GNP (towards the target value for the steady solution), $\Delta T(t, r = 0)$, are shown in Fig. 9.6 (in the linear time scale) and in Fig. 9.7 (in the logarithmic scale). The heating phase was set on until the ΔT value

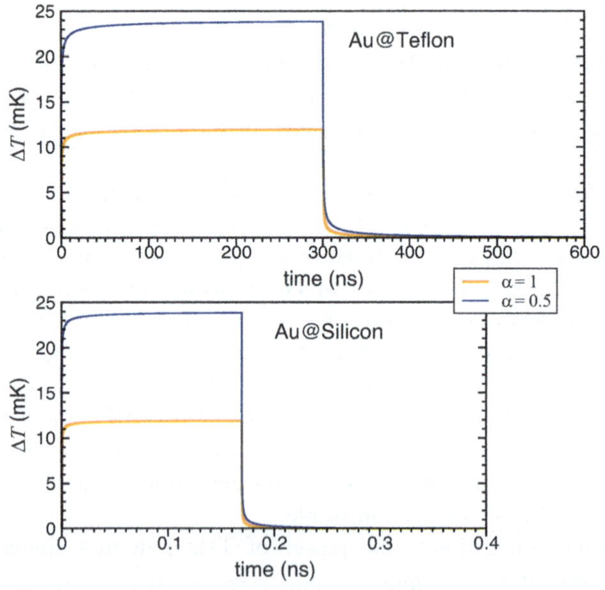

Fig. 9.6 Temperature rise ΔT versus heating/cooling time t for a GNP of radius $R_0 = 4.25$ nm in Teflon® and silicon spherical shells for two values of Q_T (for Teflon®) and Q_{Si} (for silicon) from Table 9.4. Upper curves correspond to emissivity factor of GNPs $\alpha = 0.5$, lower curves – to $\alpha = 1$

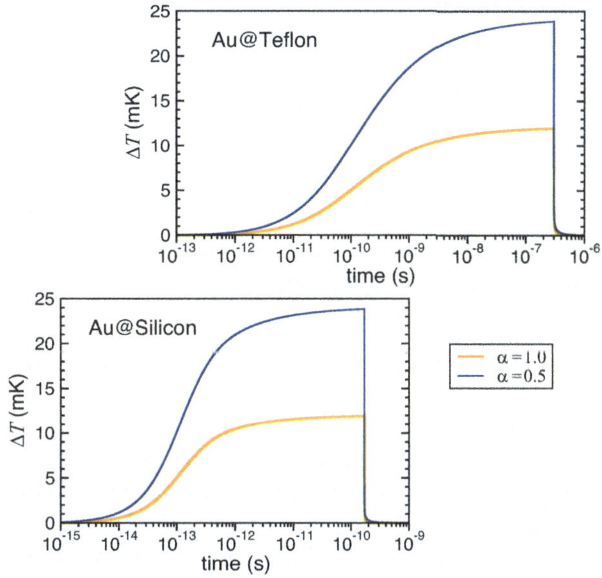

Fig. 9.7 Similar to Fig. 9.6, using the logarithmic time scale

reached an (arbitrarily chosen) level of 99.5% of the saturation value, after which the heating was set off. The resulting "response functions" $\Delta T(t)$ provide a measure of the detector's characteristic reaction time. In silicon, thermal conductivity is much higher than that in Teflon® (see Table 9.3). Since $Q_{\rm Si} \gg Q_{\rm T}$, the heating/cooling times of 8.5 nm diameter GNPs in silicon matrix are much shorter than those in Teflon® matrix. In the logarithmic scale (Fig. 9.7), one can easily see that the characteristic time at which half of the target temperature is attained is by three orders of magnitude longer in the Teflon® matrix ($\sim 10^{-10}$ s) than in silicon ($\sim 10^{-13}$ s). However, even for the Teflon®-based converter the response time is likely to be well below microseconds and hence sufficient for real-time imaging of tissue or patient. The "bonus" of the "slow" Teflon® matrix is that much lower THz power levels are required to achieve comparable heating.

Because of its very short heating/cooling time, the THz-to-IR converter on the GNP/silicon basis might be used for registering fast processes in the course of which an appreciable power of THz radiation is released. For medical imaging, on the contrary, the GNP/Teflon® converter seems to be advantageous, since it would afford an acceptable registration rate even at THz powers two orders of magnitude lower than those needed in combination with the silicon matrix.

As a general observation based on a number of trial calculations for a nanoparticle immersed in the matrix of as low thermal conductivity as that of Teflon®, the exact choice of transport parameters for the nanoparticle (confinement-dependent and estimated for a given particle size vs bulk) becomes relatively irrelevant. As is evident from Eq. (9.6), the temperature evolution would follow the uniform scaling of the heat source term, therefore the response time wouldn't depend on the emissivity factor of individual particles, but only on the GNP size and the embedding material. The temperature sensitivity of the IR camera may be a limiting factor, though.

9.4.4 On enhancement of Conversion Efficiency of THz Radiation into Heat in GNPs

We selected gold as promising metal for nanoparticles because it does not oxidise at room temperatures, hence, surfaces of GNPs are clear of oxides, which could absorb THz radiation. However, in gold, a number of Fermi electrons able to absorb the THz photons is not large. We think, it is possible to enhance the number of electrons at the Fermi level of gold by two ways: (a) by introducing impurity atoms like Fe or Ta, which form electron *d* states at the Fermi level of gold [38]; (b) by alloying gold with palladium, which possesses peak of electron DOS at the Fermi level [39] and form continuous row of solid solutions with gold [40]. Palladium, like gold, does not oxidise in air at room temperatures. The alloys with compositions close to ~50 at.%Au – 50 at.%Pd seem promising in the sense of being the most distorted, because they are "as much as possible away" from lattices of pure constituents. The distortion induces intense scattering of electrons, which may absorb the energy of THz photons and channel this energy into lattice vibrations. Consequently, the conversion of THz radiation into heat is expected to be enhanced by alloying. Indeed, the de Broglie wavelength of the Fermi electrons (as well as electrons, which absorbed the THz photons) is ~0.5 nm. This happens to be reasonably commensurate with the expected areas with distorted periodicity in the alloy made out of Au and Pd, characterized by lattice parameters of 0.408 and 0.389 nm, respectively [40]. The following observations summarize the arguments as for the "usefulness" of alloying:

(i) as compared to pure gold, the Au-Pd alloys possess enhanced electron DOS at the Fermi level [41–43]; in these conditions, due to a thermal smearing of the DOS distribution, the electron states near to the Fermi level are occupied partially. Therefore, in Au-Pd alloys, on the one hand, a number of Fermi electrons-absorbers of THz photons is larger than in pure gold, and on the other hand, an enhanced number of unoccupied electron states close by the Fermi level causes an intensive scattering of excited electrons (which absorbed THz photons);
(ii) the specific electrical resistivity of the Au-Pd alloys are several times higher than that of pure gold; moreover, at compositions ~50 at.%Au – 50 at.%Pd, it reaches maximum values and one order of value higher than that of pure gold [44, 45].

9.5 Discussion

When operating the schemes shown in Figs. 9.1 and 9.2, the previously suggested THz emitters [23, 24] and the present THz-to-IR converter are intended to be used in tandem. The advantage would be that the GNPs in the THz-to-IR converter would absorb THz photons emitted by gold nanoobjects of the emitter, i.e. photons of the same energies (1.57 or 36.0 meV). This absorption (by the Fermi electrons of 8.5 nm diameter GNPs) would happen resonantly and directly, i.e. without contributions of LAVMs considered in Ref. [26].

Estimated length × width sizes of both Teflon® and silicon matrices are 9.6 mm × 7.7 mm. Thermal conductivity of silicon is much higher than that of Teflon®, therefore, in Table 9.4, the powers Q_{Si} are much larger than Q_{T}, which results in shorter heating/cooling times as compared to those in Teflon® matrix-based the THz-to-IR converter (see Figs. 9.6 and 9.7).

Estimated heating/cooling times of 8.5 nm diameter GNPs (~169 ps/1.16 ns and ~300 ns/1.25 μs in, respectively, silicon and Teflon® matrices) show that the considered THz-to-IR converters would response to any object's evolution instantly, for all practical applications envisaged. For technical operation, either a pulsed THz radiation source (the pulse duration ≥169 ps for silicon matrix and ≥300 ns for Teflon® matrix, with pauses between pulses longer than, correspondingly, 1.16 ns and 1.25 μs), or a continuous THz radiation source would be required. In this context, the use of the THz radiation sources described in Ref. [23, 24] would be a good choice. The conventional powers required for operating the THz-to-IR converters built around the Teflon® and silicon matrices are correspondingly about 110 μW and 32.3 mW, that is, within accessible limits of contemporary THz radiation sources [46].

Kuznetsov et al. [47–53] used the THz-to-IR converter with a topological pattern of split-ring resonators. Its drawbacks are the so-called "comet tail" and image blooming effects, which deteriorate the converter's response time and spatial resolution. No such effects are likely to occur in the THz-to-IR converter based on GNPs. Indeed, in the Teflon® and silicon matrices, the edge of a reference cuboid accounted for one GNP is equal to

$$(N_{\mathrm{T}}^{*})^{-1/3} = (N_{\mathrm{Si}}^{*})^{-1/3} = (4.44 \cdot 10^{4}\ \mathrm{mm}^{-3})^{-1/3} \approx 2.8 \cdot 10^{4}\ \mathrm{nm},$$

therefore, the distance $0.5 \cdot (N_{\mathrm{T}}^{*})^{-1/3} = 0.5 \cdot (N_{\mathrm{Si}}^{*})^{-1/3} = 1.4 \cdot 10^{4}$ nm exceeds by far the maximum radii of radial distributions of temperature around heated GNPs shown in Fig. 9.6. Hence, the temperature fields of neighboring heated GNPs would not overlap. Moreover, the cooling times of GNPs are quite small. Therefore, no effects like the "comet tail" or the image blooming are a priori expected.

Characteristics of the FPA imaging devices are being improved, new types of FPAs are now making their appearance with temperature sensitivity $\Delta T_{bb} = 10$ mK

[54] and pixel's size $d = 10\,\mu m$ [55]. With this, the power sensitivities of the THz-to-IR converters would be correspondingly enhanced.

We have patented our approaches to conversion of THz vibrations into THz electromagnetic radiation [56], the THz-to-IR converter [57], as well as the source of THz radiation [58]. Also, we have published our papers [24, 59] related to these issues and now hope that these works would promote introducing the THz imaging method of the human skin cancer to the modern medical practice.

9.6 Conclusion

Contrary to ionizing X-rays, the THz radiation is a non-ionizing one and hence not harmful to living organisms. The THz medical imaging is expected to become a promising non-invasive technique for monitoring the human skin's status and for early detection of pathological conditions. In this context, the development of terahertz imaging setups emerged as one of the "hottest" areas in nanotechnology-supported modalities for the cancer diagnostics. The THz-assisted diagnostics in reflection geometry allows for non-invasive (in vivo) THz imaging of skin cancer's surface features. It could be performed in situ, without time losses on standard in vitro histological tests, which require more time. In the reflection geometry, one could investigate only skin's surface features and depth information, because the THz radiation is strongly absorbed by water and does not penetrate tissue to any significant depth. In the transmission geometry, one could study the clinically prepared tissue samples both of inner organs and the skin.

Other areas of dermatology, where the THz imaging in reflection geometry might be advantageous, are the following: skin burns and wound inspection through bandages; monitoring the treatment of skin conditions (like psoriasis), since this allows to avoid the direct contact with the skin as e.g. under ultrasound investigation. These advantages are moreover supported by the fact that the THz imaging is cheaper than the magnetic resonance one.

Summarizing, the currently available offer of highly sensitive infrared cameras allows to design schemes for the THz medical imaging of malignant tumours in biological tissue both in vitro and in vivo. The schemes contain two main parts, the terahertz-to-infrared converter and the infrared thermal imaging camera. The obtained theoretical results demonstrate that the suggested approach can be realized with the THz-to-IR converter made of Teflon® film of ~0.1 nm thickness, containing gold nanoparticles, ideally of 8.5 nm diameter. In order for this concept to be embodied, an elaboration of the process for making the Teflon® film containing gold nanoparticles is yet required. The principles applied when designing this THz-to-IR converter could be also useful in development of other devices for screening goods, non-destructive monitoring quality control, etc.

References

1. Hu BB, Nuss MC (1995) Opt Lett 20(16):1716. https://doi.org/10.1364/OL.20.001716. http://ol.osa.org/abstract.cfm?URI=ol-20-16-1716
2. Ross KFA, Gordon RE (1982) J Microsc 128(1):7. https://doi.org/10.1111/j.1365-2818.1982.tb00433.x
3. Chen JH, Avram HE, Crooks LE, Arakawa M, Kaufman L, Brito AC (1992) Radiology 184(2):427. PMID: 1620841. https://doi.org/10.1148/radiology.184.2.1620841
4. Berry E, Walker GC, Fitzgerald AJ, Zinov'ev NN, Chamberlain M, Smye SW, Miles RE, Smith MA (2003) J Laser Appl 15(3):192. https://doi.org/10.2351/1.1585079. http://scitation.aip.org/content/lia/journal/jla/15/3/10.2351/1.1585079
5. Rønne C, Thrane L, Åstrand PO, Wallqvist A, Mikkelsen KV, Keiding SR (1997) J Chem Phys 107(14):5319. https://doi.org/10.1063/1.474242. http://scitation.aip.org/content/aip/journal/jcp/107/14/10.1063/1.474242
6. Son JH (2009) J Appl Phys 105(10):102033. https://doi.org/10.1063/1.3116140
7. Oh SJ, Maeng I, Shin HJ, Lee J, Kang J, Haam S, Huh YM, Suck Suh J, Hiuk Son J (2008) In: 2008 33rd International Conference on Infrared, Millimeter and Terahertz Waves, pp 1–2. https://doi.org/10.1109/ICIMW.2008.4665813
8. Oh SJ, Kang J, Maeng I, Suh JS, Huh YM, Haam S, Son JH (2009) Opt Express 17(5):3469. https://doi.org/10.1364/OE.17.003469. http://www.opticsexpress.org/abstract.cfm?URI=oe-17-5-3469
9. Oh SJ, Choi J, Maeng I, Suh JS, Huh YM, Haam S, Son JH (2010) In: 2010 IEEE Photonics Society Winter Topicals Meeting Series (WTM), p 52. https://doi.org/10.1109/PHOTWTM.2010.5421967
10. Oh SJ, Choi J, Maeng I, Park JY, Lee K, Huh YM, Suh JS, Haam S, Son JH (2011) Opt Express 19(5):4009. https://doi.org/10.1364/OE.19.004009. http://www.opticsexpress.org/abstract.cfm?URI=oe-19-5-4009
11. Oh SJ, Huh YM, Suh JS, Choi J, Haam S, Son JH (2012) J Infrared Millimeter Terahertz Waves 33(1):74. https://doi.org/10.1007/s10762-011-9847-9.
12. Loo C, Lowery A, Halas N, West J, Drezek R (2005) Nano Lett 5(4):709. https://doi.org/10.1021/nl050127s
13. El-Sayed IH, Huang X, El-Sayed MA (2006) Cancer Lett 239(1):129. https://doi.org/10.1016/j.canlet.2005.07.035. http://www.sciencedirect.com/science/article/pii/S0304383505007378
14. Huang X, El-Sayed IH, Qian W, El-Sayed MA (2006) J Am Chem Soc 128(6):2115. https://doi.org/10.1021/ja057254a
15. Woodward RM, Wallace VP, Cole BE, Pye RJ, Arnone DD, Linfield EH, Pepper M (2002a) In: Proceedings of the SPIE 4625, Clinical Diagnostic Systems: Technologies and Instrumentation, p 160. https://doi.org/10.1117/12.469785
16. Woodward RM, Cole BE, Wallace VP, Pye RJ, Arnone DD, Linfield EH, Pepper M (2002b) Phys Med Biol 47(21):3853. http://stacks.iop.org/0031-9155/47/i=21/a=325
17. Woodward RM, Wallace VP, Pye RJ, Cole BE, Arnone DD, Linfield EH, Pepper M (2003) J Invest Dermatol 120(1):72. https://doi.org/10.1046/j.1523-1747.2003.12013.x
18. Wallace VP, Fitzgerald AJ, Shankar S, Flanagan N, Pye R, Cluff J, Arnone DD (2004) Br J Dermatol 151(2):424. https://doi.org/10.1111/j.1365-2133.2004.06129.x. https://onlinelibrary.wiley.com/doi/abs/10.1111/j.1365-2133.2004.06129.x
19. Wallace VP, Fitzgerald AJ, Pickwell E, Pye RJ, Taday PF, Flanagan N, Ha T (2006) Appl Spectrosc 60(10):1127. https://doi.org/10.1366/000370206778664635
20. Pickwell E, Wallace VP (2006) J Phys D Appl Phys 39(17):R301. http://stacks.iop.org/0022-3727/39/i=17/a=R01
21. Infrared Cameras Inc. (2018) Mirage 640 P-series | fixed/process control calibrated thermal camera with temperature measurement. https://infraredcameras.com/thermal-infrared-products/mirage-640-p-series/. Accessed 6 Aug 2018

22. FLIR A6700sc MWIR (2018) Science-grade MWIR INSB camera. https://www.flir.com/products/a6700sc-mwir/. Accessed 6 Aug 2018
23. Moldosanov K, Postnikov A (2016) Beilstein J Nanotechnol 7:983. https://doi.org/10.3762/bjnano.7.90
24. Postnikov AV, Moldosanov KA (2018) Nanotechnology 29(28):285704. http://stacks.iop.org/0957-4484/29/i=28/a=285704
25. TYDEX® THz materials (2018). http://www.tydexoptics.com/products/thz_optics/thz_materials/. Accessed 6 Aug 2018
26. Postnikov A, Moldosanov K (2016) In: Maffucci A, Maksimenko SA (eds) Fundamental and applied nano-electromagnetics. The NATO science for peace and security programme, Series B: physics and biophysics. Springer, Dordrecht, pp 171–201. https://doi.org/10.1007/978-94-017-7478-9. Proceedings of the NATO Advanced Research Workshop on Fundamental and Applied Electromagnetics, Minsk, Belarus, 25–27 May 2015
27. TYDEX® THz lenses (2018). http://www.tydexoptics.com/products/thz_optics/thz_lens/. Accessed 6 Aug 2018
28. Kubo R (1962) J Phys Soc Jpn 17(6):975. https://doi.org/10.1143/JPSJ.17.975.
29. Kubo R (1977) J Phys Colloq 38(C2):C2. https://doi.org/10.1051/jphyscol:1977214. https://hal.archives-ouvertes.fr/jpa-00217053
30. Moldosanov K, Postnikov A (2018) On the plausible nature of the size effect in heterogeneous catalysis on gold nanoparticles. https://arxiv.org/abs/1808.10607
31. Zhang ZM (2007) Nano/microscale heat transfer. McGraw Hill professional. McGraw-Hill Education. https://books.google.fr/books?id=64ygtm0HWtcC
32. Ashcroft NW, Mermin ND (1976) Solid state physics. Saunders College. https://books.google.fr/books?id=FRZRAAAAMAAJ
33. Lynn JW, Smith HG, Nicklow RM (1973) Phys Rev B 8:3493. https://doi.org/10.1103/PhysRevB.8.3493. http://link.aps.org/doi/10.1103/PhysRevB.8.3493
34. Moldosanov KA, Lelevkin VM, Kozlov PV, Kaveev AK (2012) J Nanophotonics 6:061716. https://doi.org/10.1117/1.JNP.6.061716
35. Gafner YY, Gafner SL, Zamulin S, Redel LV, Baidyshev VS (2015) Physics of Metals and Metallography 116(6):568. https://doi.org/10.1134/S0031918X15040055. Original Russian text published in: Fizika Metallov i Metallovedenie 116(6):602–609, 2015
36. Stojanovic N, Maithripala DHS, Berg JM, Holtz M (2010) Phys Rev B 82:075418. https://doi.org/10.1103/PhysRevB.82.075418. https://link.aps.org/doi/10.1103/PhysRevB.82.075418
37. Huang CL, Feng YH, Zhang XX, Li J, Wang G, Chou AH (2013) Acta Physica Sinica 62(2):026501. https://doi.org/10.7498/aps.62.026501
38. Postnikov AV, Moldosanov KA (2012) J Nanophotonics 6:061709. https://doi.org/10.1117/1.JNP.6.061709.
39. Christensen NE (1976) Phys Rev B 14:3446. https://doi.org/10.1103/PhysRevB.14.3446. https://link.aps.org/doi/10.1103/PhysRevB.14.3446
40. Okamoto H, Massalski TB (1985) Bull Alloy Phase Diagr 6(3):229. https://doi.org/10.1007/BF02880404
41. Lesiak B, Jozwik A (2004) Surf Interface Anal 36(8):793. https://doi.org/10.1002/sia.1766. https://onlinelibrary.wiley.com/doi/abs/10.1002/sia.1766
42. Nahm TU, Jung R, Kim JY, Park WG, Oh SJ, Park JH, Allen JW, Chung SM, Lee YS, Whang CN (1998) Phys Rev B 58:9817. https://doi.org/10.1103/PhysRevB.58.9817. https://link.aps.org/doi/10.1103/PhysRevB.58.9817
43. Oh SJ, Nahm TU (1996) J Electron Spectrosc Relat Phenom 78:43. https://doi.org/10.1016/S0368-2048(96)80023-5. http://www.sciencedirect.com/science/article/pii/S0368204896800235
44. Darling AS (1972) Gold Bull 5(4):74. https://doi.org/10.1007/BF03215168
45. Rowland T, Cusack NE, Ross RG (1974) J Phys F Metal Phys 4(12):2189. http://stacks.iop.org/0305-4608/4/i=12/a=015

46. Gallerano GP, Biedron S (2004) In Proceedings of the 2004 FEL Conference, p 216. http://accelconf.web.cern.ch/AccelConf/f04/papers/FRBIS02/FRBIS02.PDF
47. Kuznetsov SA, Paulish AG, Gelfand AV, Lazorskiy PA, Fedorinin VN (2011) Appl Phys Lett 99(2):023501. https://doi.org/10.1063/1.3607474
48. Kuznetsov SA, Paulish AG, Gelfand AV, Lazorskiy PA, Fedorinin VN, Arzhannikov AV (2011) Tech Mess 78(11):526. doi:10.1524/teme.2011.0208. https://doi.org/10.1524/teme.2011.0208
49. Kuznetsov SA, Paulish AG, Gelfand AV, Lazorskiy PA, Fedorinin VN (2012) Prog Electromagn Res 122:93. https://doi.org/10.2528/PIER11101401. http://www.jpier.org/PIER/pier.php?paper=11101401
50. Kuznetsov SA, Paulish AG, Gelfand AV, Astafiev MA, Arzhannikov AV, Fedorinin VN, Thumm MKA (2012) Proc SPIE 8423:8423. https://doi.org/10.1117/12.922728
51. Zagubisalo PS, Paulish AG, Kuznetsov SA (2014) J Phys Conf Ser 490(1):012174. http://stacks.iop.org/1742-6596/490/i=1/a=012174
52. Paulish AG, Kuznetsov SA (2016) Tech Phys Lett 42(11):1130. https://doi.org/10.1134/S1063785016110195. Published in: Pis'ma v Zhurnal Tekhnicheskoi Fiziki 42(22):64–71, 2016
53. Kuznetsov SA, Paulish AG, Navarro-Cía M, Arzhannikov AV (2016) Sci Rep 6:21079. https://doi.org/10.1038/srep21079. https://www.nature.com/articles/srep21079
54. DALI D900-series – cooled FPA module (2018). http://www.dali-tech.us/products/d900-series-68.html. Accessed 6 Aug 2018
55. Sofradir DAPHNIS-HD MWIR detector (2018). http://www.sofradir.com/product/daphnis-hd-mw/. Accessed 6 Aug 2018
56. Moldosanov KA, Postnikov AV (2018) Converter of terahertz vibrations into terahertz electromagnetic radiation. Russian patent RU 2650343. Priority: 20.03.2017, date of publication: 11.04.2018 (Bull. 11). http://www1.fips.ru/fips_servl/fips_servlet?DB=RUPAT&DocNumber=2650343&TypeFile=html (2017). https://patents.google.com/patent/RU2650343C1/en. Accessed August 7, 2018
57. Moldosanov KA, Lelevkin VM, Kairyev NZh, Postnikov AV (2016) Terahertz-infrared converter for visualiation of sources of terahertz radiation. Russian patent RU 2642119. Priority: 21.06.2016, date of publication: 24.01.2018 (Bull. 3). http://www1.fips.ru/fips_servl/fips_servlet?DB=RUPAT&DocNumber=2642119&TypeFile=html. https://patents.google.com/patent/RU2642119C2/en. Accessed Aug 7 2018
58. Moldosanov K, Postnikov AV (2018) Source of terahertz radiation. Russian patent RU 2622093. Priority: 13.05.2016, date of publication: 09.06.2017 (Bull. 16). http://www1.fips.ru/fips_servl/fips_servlet?DB=RUPAT&DocNumber=2622093&TypeFile=html (2016). https://patents.google.com/patent/RU2622093C1/en. Accessed 7 Aug 2018
59. Moldosanov KA, Postnikov AV, Lelevkin VM, Kairyev NJ (2017) Ferroelectrics 509(1):158 (2017). https://doi.org/10.1080/00150193.2017.1296344.

Chapter 10
Carbon-Based Terahertz Resonant Antennas

Antonio Maffucci and Sergey A. Maksimenko

Abstract Given their fascinating properties, carbon nanomaterials are currently proposed for the realization of devices working in the terahertz range, overcoming the limits of the conventional materials. Several types of devices have been proposed, designed, and theoretically studied, based on different physical phenomena observed in such materials. In this Chapter, we review the state of the art of the study on THz resonant antennas, based on features of the propagation of the surface plasmon polaritons.

Keywords Carbon nanotubes · Graphene · Nanoelectronics · Nanoelectromagnetics · Plasmon resonances · Terahertz range

10.1 Introduction

A challenging frontier in electronics and electromagnetics is the Terahertz technology, operating in the frequency range of 0.1–10 THz, also known as the submillimeter wave range (1–0.03 mm). The increasing interest in this technology is related to the potential applications of THz spectroscopy [1–4].

The THz range is a "boundary" region where the classical electronics or photonics technologies fail. Classical electronics devices exhibit a dramatic drop of performance for frequencies above 0.1 THz, due to the degradation of the transport properties. Indeed, it is extremely difficult to realize a compact THz

A. Maffucci (✉)
Department of Electrical and Information Engineering, University of Cassino and Southern Lazio, Cassino, Italy

INFN – LNF, Frascati, Italy
e-mail: maffucci@unicas.it

S. A. Maksimenko
Institute for Nuclear Problems, Belarusian State University, Minsk, Belarus

A. Maffucci, S. A. Maksimenko (eds.), *Fundamental and Applied Nano-Electromagnetics II*, NATO Science for Peace and Security Series B: Physics and Biophysics, https://doi.org/10.1007/978-94-024-1687-9_10

circuit with conventional materials and classical design approaches: it is difficult to design lumped elements whose characteristic dimensions become comparable to or higher than the wavelength. For the same reason, the electrical interconnects do no longer behave as transmission lines carrying only the fundamental mode, but they rather act as multimode channels. This makes difficult the frequency tuning, limits the range of operating frequencies, and introduces high sensitivity to errors of their setting. In addition, hollow single-mode waveguides and resonators becomes unrealistic because they would require an unaffordable surface purity, to avoid the losses due to the scattering of modes on irregular surfaces. Conversely, classical photonics devices fail for frequencies below 10 THz, due to the steep increase of the conversion of the photon energy to the thermal energy. The term "THz gap" is therefore currently adopted to indicate the lack of efficient power sources, detectors, and low-loss interconnects in such a range, compared to the well-assessed RF/microwave technology (electronics) or infrared or far infrared technology (photonics). This difficulty has meant that the THz range has only begun to be explored thoroughly over the last decade [5–7]. Despite recent advances [8], THz technology is not currently being fully-exploited due to the immature state of the devices, such as antennas, detectors, sources, amplifiers and so on. Realizing such devices is nowadays one of the most challenging tasks [9].

Addressing the THz technology has also increased the tendency of blurring the classical boundary between electronics and photonics, giving rise to mutual penetration of the correspondent concepts [10–13]. Indeed, quasi-optical principles moved to the radio-frequencies, so that even coherent lumped circuits (capacitors, inductors, resistors, interconnects) have been implemented in the optical frequency range [14].

An alternative route to THz technology has recently been opened by the impressive progress of the nanotechnology, which has led to the synthesis and fabrication of different nanostructured materials with fascinating mechanical, electronic and optical properties irreducible to that of classical bulk materials [15, 16]. Single-mode components (photonic crystals, transmission lines, microcavities, antennas) and lumped elements (capacitors, inductors, resistors, interconnects) have been fabricated by means of nano-sized elements, exploiting their novel electromagnetic properties. In general, they are related to their discrete spectrum of energy states, which is a consequence of the spatial confinement of the charge carriers motion to sizes comparable with the de Broglie wavelength. These nanostructures are characterized by novel dispersion providing their unusual electromagnetic response.

Given their outstanding electrical, thermal and mechanical properties (e.g., [17–19]), carbon-based materials, like carbon nanotubes (CNTs) or graphene nanoribbons (GNRs) have been successfully proposed for realizing devices for microwave and RF nanoelectronics (e.g., [20–26]) as well as for photonics applications (e.g., [16, 27–29])

In particular, graphene-based materials have been demonstrated to support the propagation of the so-called surface plasmon polaritons (SPPs), which are characterized by a strong wave localization and moderate loss, and whose resonances appears in the THz and Infrared (IR) frequency ranges. A really promising feature of such materials is the possibility of being tunable by means of electrical and/or magnetic bias or chemical doping. Therefore, the possible use of carbon nanomaterial in the THz range applications has been recently investigated [16, 30, 31] in view of realizing, in particular, filters [32], nanointerconnects [33, 34], ballistic transistors [35], waveguides and nanoantennas [36], and emitters [37, 38].

Despite the impressive progress in graphene technology observed in the past years, it is important to stress that, to date, carbon-based THz devices have not yet been realized. Several types of devices have been proposed, designed, and theoretically studied, but few real-world application is already available, such as the prototypes of detector described in [39], made by graphene field-effect transistors, or the recently-proposed detector made by antenna-coupled graphene transistors [40].

In this Chapter, we review the most advanced proposals for realizing THz emitters and detectors, that are based on the concept of resonant antennas, associated to the surface plasmon polaritons. After a short review on the carbon nanomaterial conductivity and on the associated frequency dispersion laws (Sect. 10.2), we will discuss the theoretical principles of SPP resonant antennas (Sect. 10.3), and review the designs proposed so far, with emphasis on both their advantages and open issues (Sect. 10.4).

10.2 Electrical Properties of Carbon Nanomaterials from DC to IR Range

Here we briefly resume the main electrical properties of two popular materials derived by the graphene and proposed as innovative conductors for novel nano-interconnects and nano-antennas in the THz range: the Carbon Nanotubes (CNTs) and the Graphene Nanoribbons (GNRs). A detailed review of the electrical properties of such materials may be found in [41, 42].

The carbon nanotubes and graphene nanoribbons are both obtained by graphene sheets: in the first case the sheet is rolled up to form a tube, whereas in the second one the sheet is simply cut into ribbons. Depending on the way the ribbons are cut and rolled-up, both CNTs and GNRs are classified in terms of their chirality, as armchair, zig-zag, or chiral, see Fig. 10.1.

The electronic band structure of the graphene is described, in the nearest-neighbors tight-binding approximation, by the energy dispersion relation of the valence electrons that are not involved in the bonds and hence are free to move (the so-called π-electrons), which can be written as [42]:

$$E^{(\pm)}(\mathbf{k}) = \pm\gamma\left[1 + 4\cos\left(\frac{\sqrt{3}k_x a_0}{2}\right)\cos\left(\frac{k_y a_0}{2}\right) + 4\cos^2\left(\frac{k_y a_0}{2}\right)\right]^{1/2}, \quad (10.1)$$

where the signs "+" and "–" indicate the conduction and valence band, respectively, and $\gamma = 2.7\text{eV}$ is the carbon-carbon interaction energy. The valence and conduction

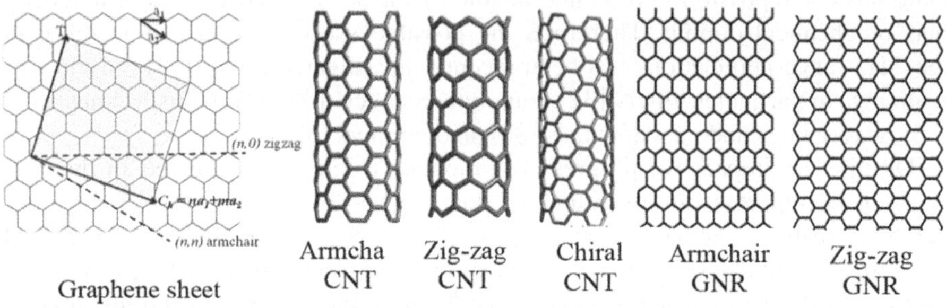

Fig. 10.1 Carbon nanotubes and graphene nanoribbons derived from a graphene sheet: different cutting directions in the graphene lattice lead to a different chirality

bands touch themselves at the so-called *Fermi points*, in whose neighborhood the dispersion relation (10.1) may be given a linear approximation as:

$$E^{(\pm)} \cong \pm \hbar v_F \left| \mathbf{k} - \mathbf{k}_0 \right|, \tag{10.2}$$

being $\mathbf{k}_0$ the wavevector at a Fermi point, v_F the Fermi velocity of the π-electrons, and $\hbar$ the Planck constant. At the equilibrium, all the π-electrons completely fill the valence band, but since between this band and the conduction one there is no gap, graphene may be considered as a conductor rather than a semiconductor. In other words, under the action of any external field the π-electrons can jump into the conduction band so producing charge motion. Graphene was proved to exhibit a very high electron mobility, up to 15000 $\text{cm}^2{\cdot}\text{V}^{-1}{\cdot}\text{s}^{-1}$.

A *carbon nanotube* (CNT) is obtained by rolling up a surface cut in a graphene layer, according to a chiral vector $\mathbf{C} = n\mathbf{a}_1 + m\mathbf{a}_2$ and a translational vector $\mathbf{T} = t_1\mathbf{a}_1 + t_2\mathbf{a}_2$, where n and m are integer indexes defining the CNT. From these indexes, for instance, it is possible to calculate the CNT circumference as $C = a_0\sqrt{n^2 + nm + m^2}$. In practical cases, the CNT length is very large compared with the length of the lattice unit, thus the longitudinal wave vector k is almost continuous. Instead, the transverse dimension (ranging from nm to tens of nm) is comparable to the de Broglie wavelength and thus the transverse wave vector $k_\perp$ is quantized: it takes only the discrete values $\mu \Delta k_\perp$ with $\mu = 0, 1, \ldots, N-1$. In the zone-folding approximation, the dispersion relation for a single wall carbon nanotube (SWCNT) is given by [43]:

$$E_\mu^{(\pm)}(k) = E^{(\pm)}\left(k\frac{\mathbf{K}_2}{|\mathbf{K}_2|} + \mu\mathbf{K}_1\right) \text{ for } -\frac{\pi}{T} < k < \frac{\pi}{T} \quad \text{and} \quad \mu = 0, 1, \ldots N-1, \tag{10.3}$$

where T is length of the translational vector $\mathbf{T}$, and the vectors $\mathbf{K}_1$ and $\mathbf{K}_2$ are given by $\mathbf{K}_1 = (-t_2\mathbf{b}_1 + t_1\mathbf{b}_2)/N$ and $\mathbf{K}_2 = (m\mathbf{b}_1 - n\mathbf{b}_2)/N$, being N the number of parallel segments lying in the first Brillouin zone. The CNT chirality is given by the values of the indexes n and m: if $n = 0$ or $m = 0$, they are called "zig-zag" (z-CNT), if $n=m$ they are denoted as "armchair" (a-CNT), otherwise they are generally indicated as "chiral". By studying the electronic band structure in all these cases, it turns out that the z-CNTs have always a zero gap (they are metallic), whereas a-CNTs may have or not a finite gap (they can be either metallic and semiconducting). Statistically, if CNTs grow without any control on their chirality, in a population of CNTs only 1/3 is metallic and the rest is semiconducting.

The graphene nanoribbons are obtained by cutting a graphene layer in ribbons of a transverse width w much smaller than the longitudinal length. Therefore, as it happens for the CNTs, the longitudinal wave vector is almost continuous, whereas the transverse one is quantized. The "armchair" (a-GNR) or "zigzag" (z-GNR) are the two basic shapes that may be obtained by changing the cutting direction (see Fig. 10.1). The width w of a GNR is directly related to the integer N, indicating the number of dimers (two carbon sites) for the a-GNRs and the number of zigzag lines for the z-GNRs.

In the nearest neighbor tight-binding approximation, the dispersion relation of an a-GNR for the generic μ-th band is given by [44, 45]:

$$E_{\mu}(k) = \gamma\sqrt{1 + 2\varepsilon_{\mu}\cos(kT/2) + \varepsilon_{\mu}^{2}} \quad \text{for} \quad -\frac{\pi}{T} \leq k \leq \frac{\pi}{T}, \tag{10.4}$$

where $\varepsilon_{\mu} = 2\cos(p_{\mu})$, being p_{μ} the transverse wavenumber. In the same conditions, the dispersion relations for z-GNRs may be obtained by solving the eigenvalue problem [44, 45]:

$$H(k)\,\mathbf{u} = E(k)\,\mathbf{u}, \tag{10.5}$$

As for the bandgap, it is always zero for z-GNRs (they are metallic), whereas it can be non-zero for z-GNRs (they can be either metallic and semiconducting). The statistical distribution of metallic and non-metallic GNRs follows the same considerations done for CNTs.

The electrical conductivity of CNT and GNR exhibits a very complex behavior in frequency. Typically, three main regions may be distinguished:

(i) a low frequency range, where a quasi-static model holds, and metallic CNTs and GNRs exhibit a classical Drude behavior;
(ii) an intermediate region where a monotonic dependence of the conductivity with the frequency can be observed;
(iii) a high frequency range, where resonances occur due to the strong effect of the interband transitions. Figure 10.6 shows an example of the computed axial conductivity for a metallic single-walled CNT of length 200 nm: in this case the low and high frequency ranges correspond to $f < 1$ THz ($\lambda > 300$ μm), and $f > 100$ THz ($\lambda < 3$ μm), respectively.

In the low frequency range (long wavelength) the axial conductivity for a CNT or a GNR may be expressed as:

$$\sigma_c(\omega) = \frac{\sigma_0}{1 + i\omega/\nu}, \tag{10.6}$$

being ν the collision frequency (the inverse of the relaxation time) and σ_0 the DC value. A general expression for such a term is the following [46, 47]:

$$\sigma_0 = \frac{2 v_F M}{\upsilon R_0 X}, \tag{10.7}$$

where $R_0 = 12.9\ \text{k}\Omega$ is the so-called *quantum resistance*, M is the *equivalent number of conducting channels*, v_F is the Fermi velocity, whereas $X = \pi D$ for a CNT of diameter D and $X = \pi W$ for a GNR of width W. The low frequency conductivity is therefore modulated by the dimension of the carbon material and by the number of channels which, in turns, depend on dimension, chirality and temperature, see [45].

In the intermediate region, the conductivity cannot be given a closed form, but it must be derived from the dispersion relation for the propagating surface wave, which for a CNT is given by [46, 47]:

$$\frac{\kappa^2}{k^2} K_q(\kappa r_c) I_q(\kappa r_c) = \frac{ic}{4\pi r_c \sigma_{zz}} \left[1 - \left(\kappa^2 + k^2\right)\right] \tag{10.8}$$

where $K_q(.)$ and $I_q(.)$ are the modified Bessel functions, c is the speed of light in vacuum, $k^2 = k^2 + h^2$, and h is the surface wave wavenumber.

In THz frequency range, a semi-classical expression for the conductivity is the following, obtained when only intraband transitions are taken into account [48]:

$$\sigma_{zz} = -i \frac{e^2 k_B T}{\pi \hbar (\omega - i\upsilon)} \ln\left\{2\left[1 + \cosh\left(\frac{\mu}{k_B T}\right)\right]\right\}, \tag{10.9}$$

where k_B and $\hbar$ are the Boltzmann and Planck constants, respectively, T is the temperature and μ the chemical potential. For higher frequencies, interband transitions occur.

A typical behavior is reported in Fig. 10.2 that shows the axial conductivity versus the wavelength, for a single-walled CNT of length 200 nm. The high-frequency range (short wavelength) exhibits the plasmon resonance, whereas the solution of (10.8) highlights an unusual electromagnetic response in the intermediate range, with an almost linear dependence of the conductivity versus the wavelength. This provides an almost frequency-independent wavenumber.

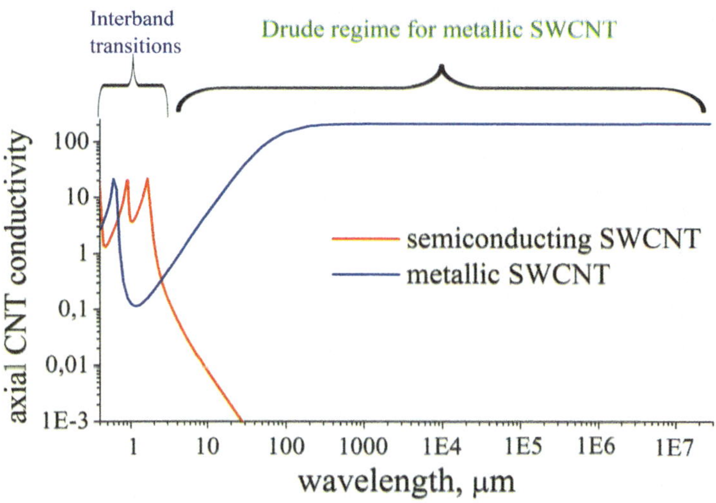

Fig. 10.2 Axial conductivity versus wavelength for a 200 nm-long SWCNT

10.3 Surface Plasmon Polaritons Terahertz Resonances

The most studied mechanism for realizing carbon-based antennas working in the THz range is the excitation of plasmon resonances, and the realization of the so-called resonant antennas. These are the most popular antennas, widely used in all applications, and their implementation with graphene materials has been studied from the earlier works on this topic, e.g. [19, 36, 46, 49].

As pointed out in the introduction, the phenomenon exploited in such antennas is the propagation of slowly decaying surface waves, the so-called surface plasmon polaritons (SPPs). Since these waves propagate with a slow phase velocity (typically, 2 orders of magnitude smaller than the light speed in vacuum), the resonances of an antenna made by graphene ribbon or carbon nanotube may be found in the THz and IR frequency ranges.

The theoretical demonstration of the SPP existence may be carried out starting from the study of the electrodynamics in carbon materials. A possible approach is for instance presented in [47] with reference to CNTs. The method of Effective Boundary Conditions is used, considering the CNT surface as a smooth homogeneous surface. The quantum nature of the CNT electrodynamics is taken into account by a rigorous evaluation of its conductivity, solving a system of kinetic equations for the density matrix which describes the dynamics of CNT electrons under the action of an external electric field, taking into account both intraband and interband transitions. By using this approach, the typical frequency behavior of $\beta = k/h$ is found to be that reported in Fig. 10.3, with a real part almost independent on frequency in the THz range. In addition, the phase velocity $\boldsymbol{v_{ph}} = \boldsymbol{\beta c}$ is much smaller than the speed of light in vacuum (about two orders of magnitude):

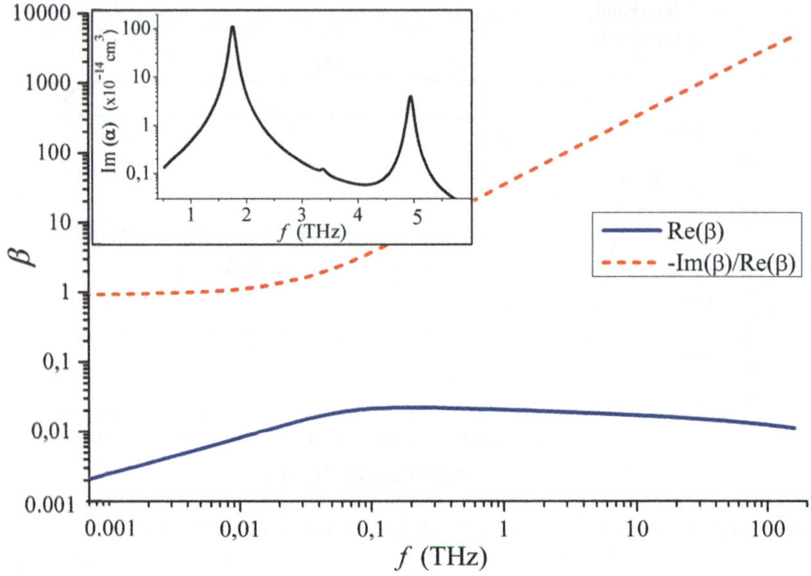

Fig. 10.3 Typical behavior of the slow-wave coefficient β versus frequency in a metallic SWCNT, with b=1.42 Å [47]. Inset: antenna resonances of a 0.6 μm-long metallic SWCNT [46]

this strong slowing-down of the waves is responsible for the possibility of having antenna resonances in the THz or IR ranges associate to the effect of finite length, as shown in the inset of Fig. 10.3, referring to a metallic zig-zag SWCNT of length 0.6 μm.

Similar results hold for a graphene sheet of conductivity $\boldsymbol{\sigma}$, where the dispersion relation of the supported SPPs may be approximated as [32]:

$$k = k_0\sqrt{1 - \left(\frac{2}{\zeta\sigma}\right)^2}, \tag{10.10}$$

being $\boldsymbol{k_0}$ the free-space wavenumber and $\boldsymbol{\zeta}$ the intrinsic impedance of the surrounding media. The above simple relation refers to infinite sheets, but can be used to provide an estimation of the SPP wavelength in realistic structures like graphene nanoribbons. Indeed, Fig. 10.4 shows the variation of the first plasmon resonance with respect to the width W and the length L of a GNR deposited on a dielectric substrate [50].

The strong slowing down of surface wave in CNTs makes them prospective for the design of IR and THz nanoantennas [19, 46, 49] (see inset in Fig. 10.3). However, the observation of the antenna resonances in an isolated CNT is a very challenging problem because of the small antenna efficiency of single-walled nanotubes, of the order of 10^{-4}–10^{-6} [19, 49] (see Sect. 10.4 for details). Even so, antenna resonances in SWCNTs can be proposed [51] as the physical mechanism

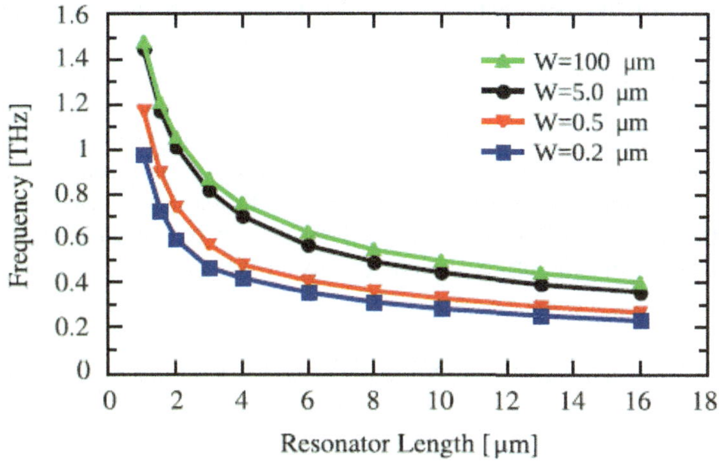

Fig. 10.4 Computed value of the first plasmon resonance for a GNR antenna, versus the length *L*, for different values of the width *W*. (Reprinted with permission from [50], © 2012 Elsevier)

of the conductivity non-Drude behavior in CNT-based composites [52–56], which exhibit a pronounced peak in the THz range. Below we briefly present a theoretical model which couples antenna resonances in isolated CNTs to the THz conductivity peak in composites materials. The model is used for the interpretation of the experimental results reported in [54].

Let us consider a dilute composite material in free space comprising randomly dispersed, randomly oriented, achiral, SWCNTs of different lengths. We designate the single index j to identify the type of SWCNT. The distribution function $N_j(L)$ describes the number density of SWCNTs of type j and length L. All orientations are equiprobable and $\boldsymbol{\rho}$ denotes the volume fraction occupied by the SWCNTs. This model requires the calculation of the complex-valued effective conductivity $\boldsymbol{\sigma}_{eff}(\boldsymbol{\omega})$ of the composite material, as function of the angular frequency $\boldsymbol{\omega}$, using two different approaches. A long-wavelength approach is valid when the length L of all SWCNTs is much smaller the free-space wavelength, while a short-wavelength approach is applicable when the length L is comparable or exceeds the wavelength and thus finite-length effects vanish. In the first case the simple Waterman-Truell formula can be adopted [57, 58] to estimate the effective refractive index of the chosen composite material as follows:

$$n_{eff}^2(\omega) \cong 1 + \frac{4\pi\rho}{3} \sum_j \int_0^\infty a_j(\omega, L)\, N_j(L) dL, \tag{10.11}$$

$$\sigma_{eff}(\omega) = \frac{\omega\left[n_{eff}^2(\omega) - 1\right]}{4\pi i}, \tag{10.12a}$$

where $\boldsymbol{\alpha}_j(\boldsymbol{\omega}, \boldsymbol{L})$ is the polarizability of CNTs. Following the theoretical model of a finite-length SWCNT [46] as an infinitely thin hollow cylinder with effective axial conductivity found from quantum mechanical theory [47] the polarizability of a finite length SWCNT has been calculated in broad frequency range starting from 10^{-4} THz and up to hundreds THz, where interband transitions come into play and where surface-wave propagation in SWCNTs is strongly attenuated (see Fig. 10.3) suppressing finite-length effects. In short-wavelength approach, the integral equation for scattering by a SWCNT can be solved using the Born approximation [46]. The effective conductivity of the composite material is then given by

$$\sigma_{eff}(\omega) \cong \frac{2\pi}{3} \sum_j R_j \sigma_z^{(j)}(\omega) \int_0^\infty N_j(L) dL, \tag{10.12b}$$

where $\boldsymbol{\sigma}_z^{(j)}(\boldsymbol{\omega})$is the axial conductivity of the j-th CNT.

Combining both approaches it is possible to calculate the effective conductivity of the composite: Fig. 10.5 demonstrates the spectra of the real part of effective conductivity of SWCNT-based composite material in the broad frequency range from ultraviolet to radiofrequency range. The spectra can be divided into three parts. High frequency part (f > 100 THz) is determined by interband electron transitions – in this range effective conductivity does not depend on SWCNT length. The range of antenna resonances, (1–100 THz), has peaks associated with antenna effect in SWCNTs. The most intensive peak is a first antenna (or localized plasmon)

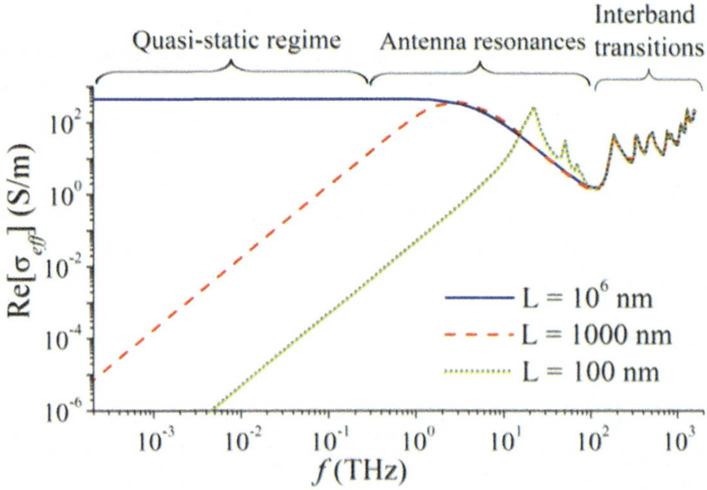

Fig. 10.5 Frequency dependence of the effective conductivity of SWCNT composite material comprising of identical SWCNTs of lengths 100 nm (dotted line), 1000 nm (dashed line), and 106 nm (solid line). SWCNT volume fraction is 1.3%; host matrix was assumed to be air [51]

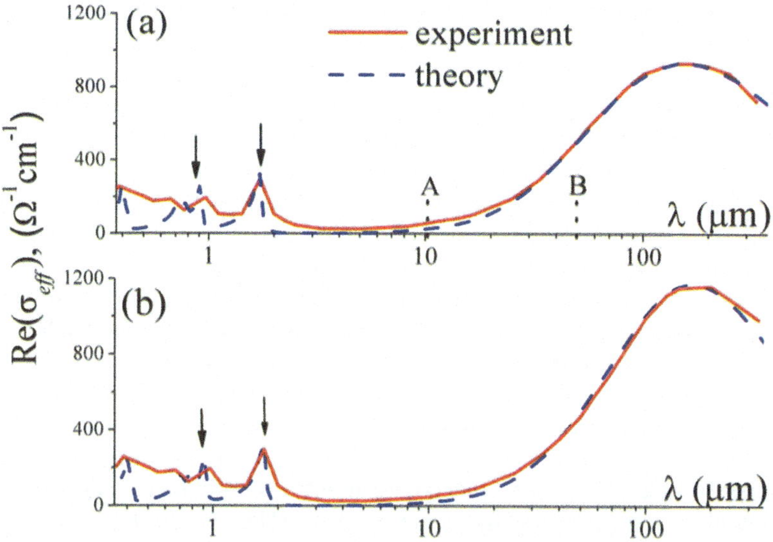

Fig. 10.6 Variations in **Re**(σ_{eff}) with **λ** at (*a*) T =300 K and (*b*) T=50 K. Solid lines: experimental data from Fig. 10.3 of Ref. [54]. Dashed lines: theoretical results. The terahertz conductivity peak is on the right in both figures. Arrows indicate the resonance lines due to the first two optical transitions of semiconducting SWCNTs (Reprinted with permission from [51] © 2018 American Physical Society)

resonance, see inset in Fig. 10.3. The low frequency range below antenna resonances (f < 1 THz) corresponds to quasi-static regime; in this regime the real part of carbon nanotube polarizability equals to SWCNT static polarizability. In quasi-static regime strong depolarizing field suppresses current excitation in finite length SWCNT resulting is small effective conductivity of the composite media (compare curves in Fig. 10.5 for longest and shortest tubes).

Figure 10.6 shows that this theoretical model comprising long- and short-wavelength approaches, as appropriate, for different spectral regimes describes very well the form of the experimentally observed terahertz conductivity peak. For this comparison, we assumed that $N_j\,(L) \sim \exp\left[-(L - L_0)^2/\Delta\sqrt{2}\right]$ is Gaussian and independent of j, with L_0 = 1.6 μm and Δ = 0.6 μm. These values were chosen to match the frequency, line width, and amplitude of the terahertz peak in the experimental data of [54].

Besides, as one can see, the predicted amplitudes of resonance lines due to first two optical transitions of the semiconducting SWCNTs coincide reasonably well with the experimental values. This coincidence is remarkable because L_0 and Δ has been chosen only to match the experimentally observed terahertz peak. The bandwidth of the peak is determined with high accuracy by electron relaxation processes (relaxation linewidth 20 meV) and inhomogeneous broadening due to CNT length dispersion in composite materials (3 meV).

Experimental verification of the hypothesis on the antenna resonance origin of the terahertz conductivity peak has been reported for the first time in [59], where experiments show that the frequency of the conductivity peak depends on the average length of CNTs in samples and thus is unambiguously attributable to the excitation of antenna resonances, what is agrees well with calculations for a relevant physical model demonstrated by Fig. 10.6 and corresponding results in [59]. Thus, one can conclude that the experimentally observable terahertz conductivity peak in CNT-based composites [52–56] proves the strong plasmon slowing-down in CNTs and strong red shift of antenna resonances in nanotubes.

This conclusion has been supported independently in alternative experiment [60], which is based on the idea that only metallic nanotubes contribute into terahertz peak if its nature is related to antenna (plasmon) resonance, while substitutional doping of CNTs increases "metallicity" of nanotubes. Theoretical model and experimental demonstration of this increase were reported in [49, 61, 62], respectively.

The plasmon resonances may be also found in a simpler way by using the Transmission Line (TL) theory. Indeed, assuming a semi-classical approach, a low bias condition and operating frequencies up to THz, the electrical propagation along conductors made by a carbon nanotube or by a graphene nanoribbon may be modeled through the TL model. The electrical current and chemical electrical potential are solutions of the TL equations with the following per-unit-length parameters [63, 64]:

$$R = \upsilon L_k, \quad L = L_k + L_m, \quad C^{-1} = C_e^{-1} + C_q^{-1}, \tag{10.13}$$

where L_m and C_e are the classical p.u.l. magnetic inductance and electrostatic capacitance, whereas L_k and C_q are the kinetic inductance and the quantum capacitance, respectively. The term L_k is associated to the kinetic energy of the conduction electrons: it adds to the classical magnetic inductance, usually hiding it, since it is 3–4 orders of magnitude higher. As a consequence of the huge kinetic inductance, the propagation velocity along the line reduces to

$$v_p \approx \sqrt{\frac{L_K}{C_e}} \ll c = \sqrt{\frac{L_m}{C_e}} \tag{10.14}$$

which is typically 2 orders of magnitude of the velocity c obtained in an interconnect made by ideal conductors. Based on this result, we can calculate the antenna resonances associated to a CNT or GNR of length L, which appear, at a first approximation, for frequencies f_n such that:

$$f_n = n\frac{v_p}{2L}, \quad n = 1, 2, 3 \ldots \tag{10.15}$$

By using (10.15) it is easy to show that nano-antennas of lengths of tens to hundreds micron may resonate in the THz range.

10.4 THz Resonant Antennas: From Basic Concepts to Advanced Designs

Let us first analyze two basic concepts of carbon resonant antennas, one made by a single carbon nanotube and the other by a ribbon of graphene. Let us consider, for instance, a simple antenna made by a metallic SWCNT of radius r_C, and length L, as in Fig. 10.7. The input impedance of such an antenna can be computed by using the above TL model, obtaining the results shown in Fig. 10.8 [65], which refers to the case of a CNT with r_C =5 nm and $L = 20$ μm. As expected, the resonance peaks appear into the THz range.

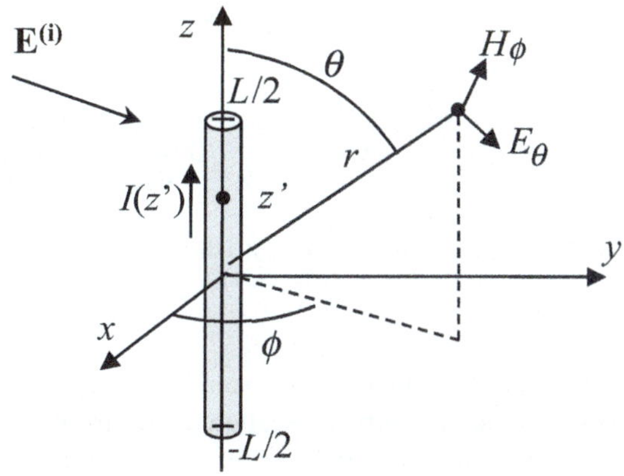

Fig. 10.7 A CNT antenna of length L, subject to an external field

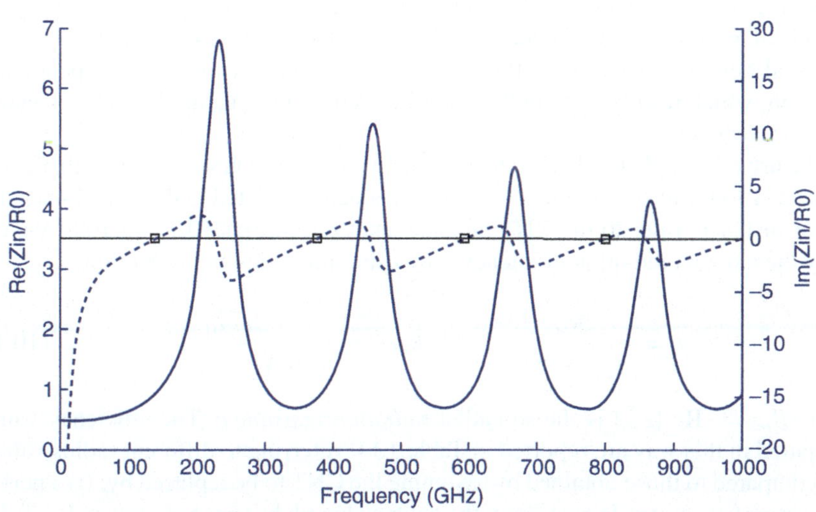

Fig. 10.8 Input impedance vs frequency for a CNT nano-antenna with r_C =5 nm and length of 20 μm, normalized to the quantum resistance $R_0 = 12.9$ kΩ [65])

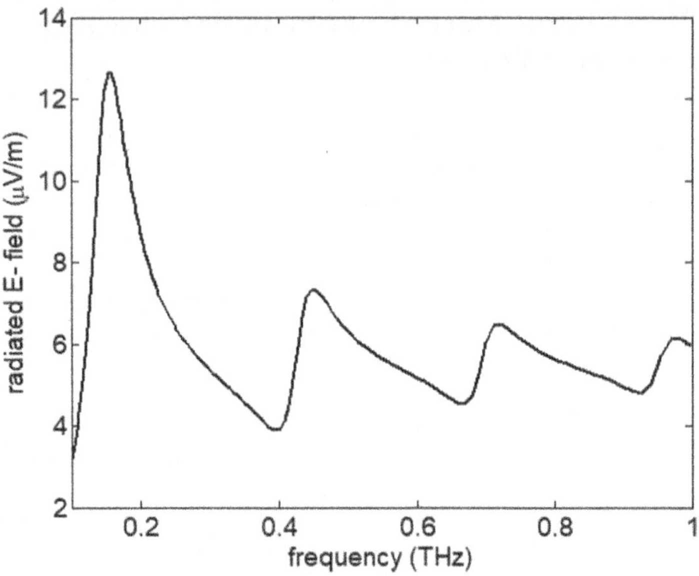

Fig. 10.9 Scattered electric field at a distance of 100 μm from a 20 μm-long CNT, with radius of 2.72 nm at 300 K, illuminated by a TEM plane wave impinging orthogonally to its axis

Let us now analyze the same structure as a detector of THz radiation. In particular, let us study the scattered field from the CNT antenna in Fig. 10.7, subject to an incident TEM field with wave-vector perpendicular to the CNT axis. The amplitude of the incident field is set to a value that excites a current along the antenna of amplitude of the order of 1 nA. Figure 10.9 shows the computed scattered electrical field at a distance of 100 μm from the antenna axis in the perpendicular direction, which results to be in the order of μV/m. Once again, the THz resonance peaks are evident.

The use of a single CNT to realize resonant antennas is however not realistic, due to the very low value of efficiency of such an antenna, related to the huge losses. Let us, for instance, refer to the CNT antenna analyzed in Figs.10.7 and 10.8. We can study the antenna radiation efficiency η by using the following definition:

$$\eta = \frac{R_{irr}}{R_{irr} + R_{HF}}, \qquad R_{HF} = \frac{l}{2\pi r_c}\sqrt{\frac{\omega\mu_0}{2\sigma}} \,. \tag{10.16}$$

where $R_{irr} =$ Re $\{Z_{in}\}$ is the so-called *radiation resistance*. The efficiency values computed in this way are reported in Table 10.1, referring to different radius values, and compared to those obtained by assuming the CNT to be replaced by: (i) a perfect conductor; (ii) copper. In each case, the antenna length has been chosen as $L = 0.5\ \lambda$. The perfect conductor case is the ideal reference case, since $R_{HF} = 0$ for $\sigma = 0$ in (10.13), hence $\eta = 1$.

Table 10.1 Radiation efficiency values for a CNT nano-antenna vs radius [65]

Conductor\radius	$r_c = 3\ \mu m$	$r_c = 30\ nm$	$r_c = 3\ nm$
Perfect conductor	$\eta = 1$	$\eta = 1$	$\eta = 1$
CNT	$\eta = 1.6{\cdot}10^{-5}$	$\eta = 1.1{\cdot}10^{-5}$	$\eta = 6.8{\cdot}10^{-6}$
Copper	$\eta = 2.9{\cdot}10^{-4}$	$\eta = 2.6{\cdot}10^{-5}$	$\eta = 7.6{\cdot}10^{-6}$

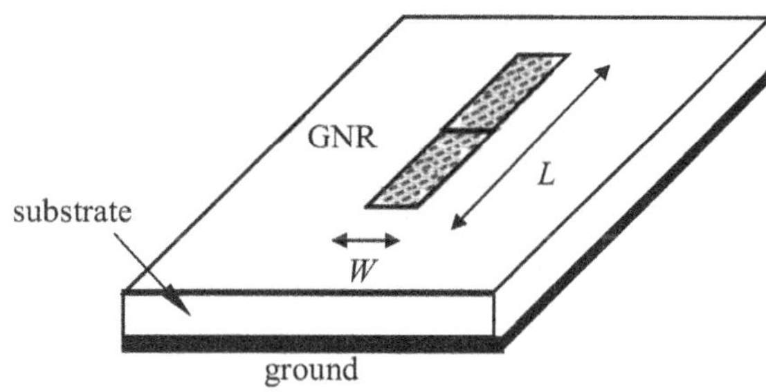

Fig. 10.10 Concept of a THz antenna realized with a pair of GNR strips excited by a copper electrode

The efficiency values for a single CNT antenna of radius of some nm are very low, and are comparable with those of a copper nano-antenna. The low values of efficiency are related to the huge resistance of a single CNT and to the steep increase of the copper resistivity when the radius scales down to nanometric dimensions. This example illustrates the recurring trade-off between miniaturization and efficiency in graphene antennas, which is the major theoretical limit on the road of the practical use of such antennas

A similar behavior is found for a nano-antenna made by a graphene nanoribbon. A conceptual antenna is described in Fig. 10.10: a metallic GNR of width W and length L is the transmitting element of a patch antenna on a dielectric substrate. A typical result for the input admittance of such an antenna is provided in Fig. 10.11, assuming a GNR width $W = 20$ nm and a total length $L = 1\ \mu$m. The result is compared to that obtained if the strips were made by copper: a copper antenna would not exhibit resonances in the THz range. The dependence of the first THz resonance frequency from W and L has been shown in Fig. 10.4. A GNR resonant antenna exhibits the same limits as the CNT one, in terms of efficiency. For instance, Fig. 10.12 shows the very low values that can be obtained for the absorption cross sections of such an antenna.

From these two conceptual examples, we can summarize that the main theoretical limit is the efficiency, so that any perspective design must address such an issue.

From the fabrication point of view, on the other side, many challenging aspects are still not solved: fabrication constraints, uncertainties, compatibility issues [66]. Indeed, despite the impressive progress and the extensive work done worldwide by several groups, the measured performance of the few available graphene structures (for instance in terms of mobility [67]) are still orders of magnitude lower than those needed to implement the ideas described here.

If losses and fabrication limits are the main drawbacks in using carbon technology for THz applications, on the other side a big advantage is given by the

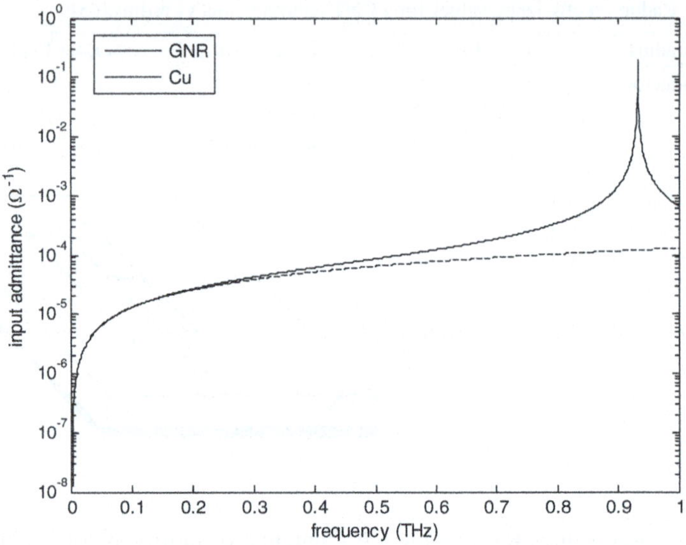

Fig. 10.11 Input admittance of a metallic graphene nanoribbon antenna of width $W = 20$ nm and length $L = 1$ μm, compared to that of a copper antenna with the same dimensions

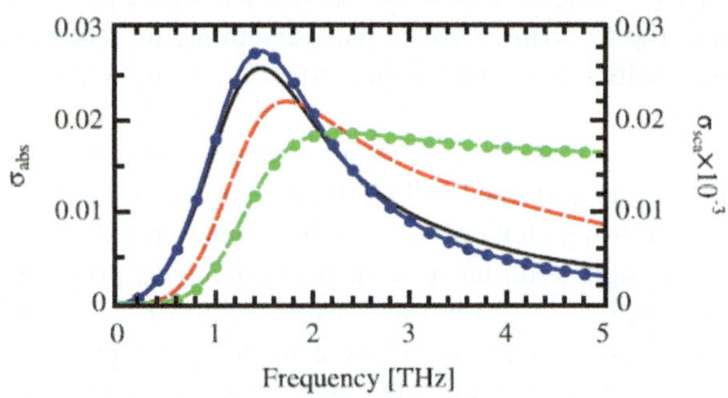

Fig. 10.12 The absorption (black line) and scattering (dashed red line) cross sections of a graphene antenna (see Fig. 10.10) on an infinite silicon substrate, obtained using numerical simulations, as compared with the absorption (blue line with dots) and scattering (dashed green line with dots) cross sections obtained using the Fabry-Perot model. (Reprinted with permission from [50], © 2012 Elsevier)

outstanding tunability properties. Indeed, it is possible to manipulate the conductivity of graphene-based materials by using external electrical or magnetic fields or by means of chemical doping. The field effect in graphene may be obtained with the simple application of an external electrostatic field: in [41], the field is perpendicular to the antenna [41], whereas in [68] it is co-planar. The external bias modify the chemical potential, which in turns affects the conductivity. This tunability have been confirmed by several experiments, also involving prototypes of devices operating from microwave range [68] to THz one [40, 69, 70].

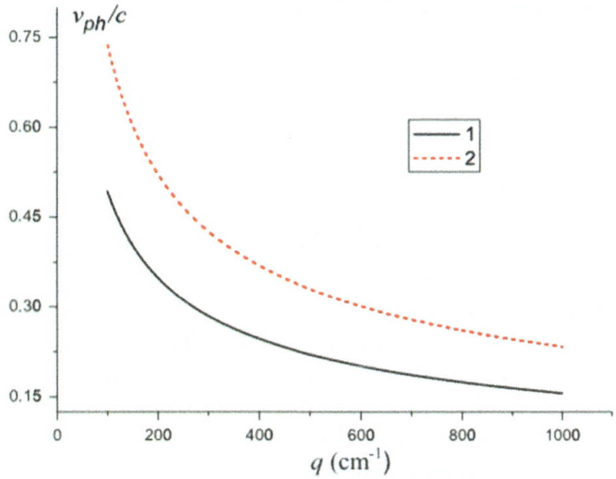

Fig. 10.13 Normalized plasmon phase velocity vs wavenumber q in a single layer graphene for two different densities of the doping electrons: (1) 10^{12} cm^{-2} and (2) 5×10^{12} cm^{-2} [62]

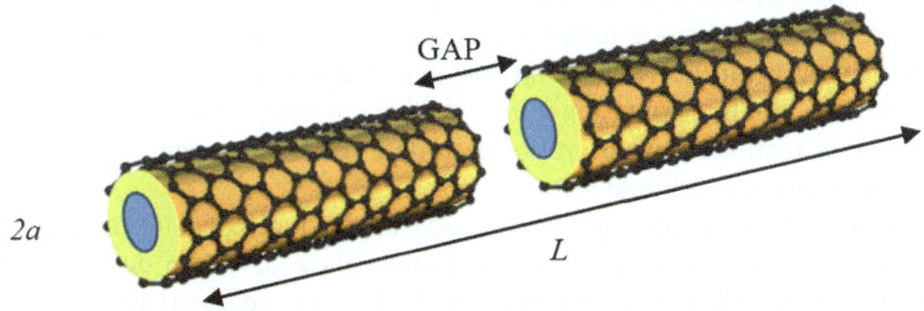

Fig. 10.14 Concept of a THz dipole antenna made by graphene coated wire

As pointed out, the position of the THz peaks in a carbon antennas can be also tuned by doping: substitutional doping has been for instance used to this end, exploiting the increase of the "metallicity" of nanotubes [61, 62].

The tunability based on the doping control may be also be observed in single-layer graphene, as shown in Fig. 10.13, where the phase velocity is shown to depend on the density of the doping [71]. It is noteworthy mentioning the possibility of using optical pumping to tune graphene's properties, as shown in [72].

An advanced design for a THz CNT antenna has been presented in [73, 74], whose concept design is reported in Fig. 10.14. Two cylindrical waveguides are used as elements for a dipole antenna. Each element is made by two coaxial cylinders of radius a (inner) and b (outer), respectively. The outer one is a dielectric, whereas the inner may be either dielectric or conductor. The wires are then surrounded by single or bi-layer graphene, so reproducing the conceptual case of a single or double-wall nanotube. The antenna can be biased either by applying electrical or magnetic field.

The better confinement properties exhibited by the proposed structure with respect to the planar counterpart allows obtaining better efficiency values in similar conditions. Indeed, by using this design for a reconfigurable THz antenna with

Table 10.2 Radiation efficiency values for the CNT antenna in Fig. 10.14 [73]

Bias voltage	CNT antenna	GNR planar antenna
(V)	η (%)	η (%)
0.46	1.2	0.16
1.84	4	0.5
7.3	14	2

length $L = 26$ μm and radius values a=450 nm, b=500 nm, the obtained value of efficiency at the first resonance are reported in Table 10.2, for different levels of the applied bias voltage. The two cylinders are made by dielectrics with permittivity $\varepsilon_r = 3.8$. The results are compared to those obtained for a GNR antenna realization (see Fig. 10.10 with the same graphene area, assuming $W = 1$ μm, and a dieletric substrate with $\varepsilon_r = 3.8$ and thickness of 1 μm. Table 10.2 shows the CNT antenna may exhibit efficiency values increased by an order of magnitude.

Graphene dipole antennas seem to be much more effective, mainly due to the better SPP confinement related to the cylindrical geometry, that do not introduce any edge effect, which is a major issue when using strips. However, from the technological point of view these antennas are much more difficult to be realized compared to the patch antennas, that have the huge inherent advantage of planarity. As a matter of fact, the design proposed to date for CNT antennas did not move significantly from this example.

Instead, the planar geometry have been strongly developed in these years, up to the level of real devices fabricated and characterized, as the detector in [40]. An advanced design has been proposed in [75] and is reported in Fig. 10.15: a planar dipole antenna is designed, where each half of the antenna consists of two by-layer graphene self-biasing patches separated by a thin Al_2O_3 layer. The lower graphene layer is in contact with the metallic electrodes of the source, that is a THz photomixer. A silicon lens is adopted to increase the directivity.

This realization demonstrates excellent properties of tunability via electrostatic doping, as shown in Fig. 10.16, where the input impedance is plotted versus frequency, parametrized to the chemical potential μ_c.

This structure conjugates the wide tunability with an outstanding stability of the impedance when the resonance frequency is shifted, as shown in Fig. 10.16. Table 10.3 shows the computed values of the radiation efficiency, that attains very high values compared to the performance of other THz antennas of the same size. The main limit is given by the strong dependence of the efficiency from the chemical potential μ_c, which is due to the effect of μ_c on the plasmon wavelength. It is also noteworthy that the use of high level of μ_c has the advantage to increase efficiency but reduces the reconfigurability.

A last example of THz antenna comes from a recently detector based on excitation of long-lived electrically tunable plasmons, see Fig. 10.16 [40].

The concept of this device is the use of coupled by-layer graphene field-effect transistors, encapsulated between two thin slabs of hexagonal boron nitride. This

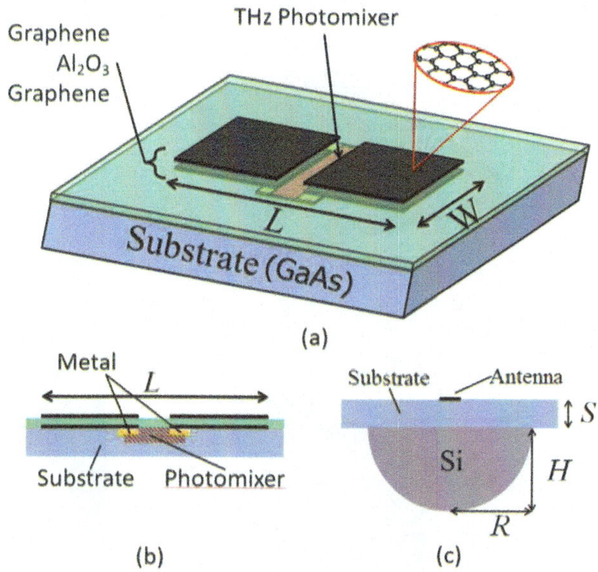

Fig. 10.15 A THz dipole antenna made by graphene: (**a**) 3D view of the structure; (**b**) cross-section of the antenna; (**c**) setup with silicon lens for better directivity. (Reprinted with permission from [75], © 2012 AIP Publishing)

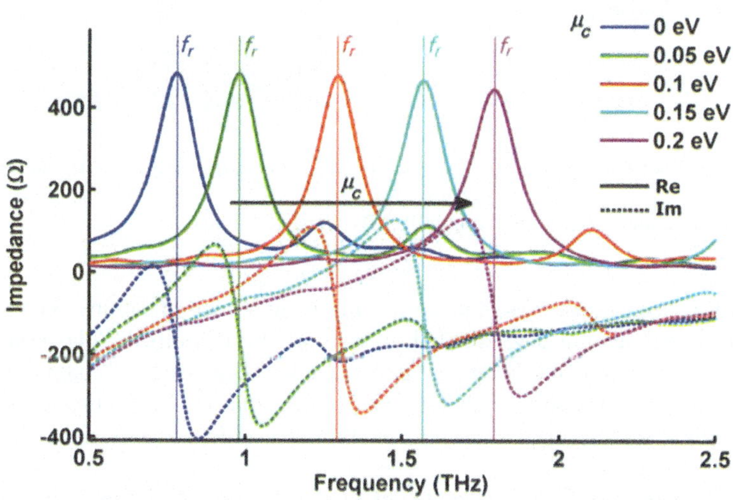

Fig. 10.16 Real and imaginary part of the input impedance of the antenna in Fig. 10.15 upon reconfiguration. (Reprinted with permission from [75], © 2012 AIP Publishing)

Table 10.3 Radiation efficiency values for the by-layer GNR antenna in Fig. 10.15

Chemical potential	Resonance frequency	η
(eV)	(THz)	(%)
0	~1	2
0.1	~1.3	7
0.2	~1.8	20

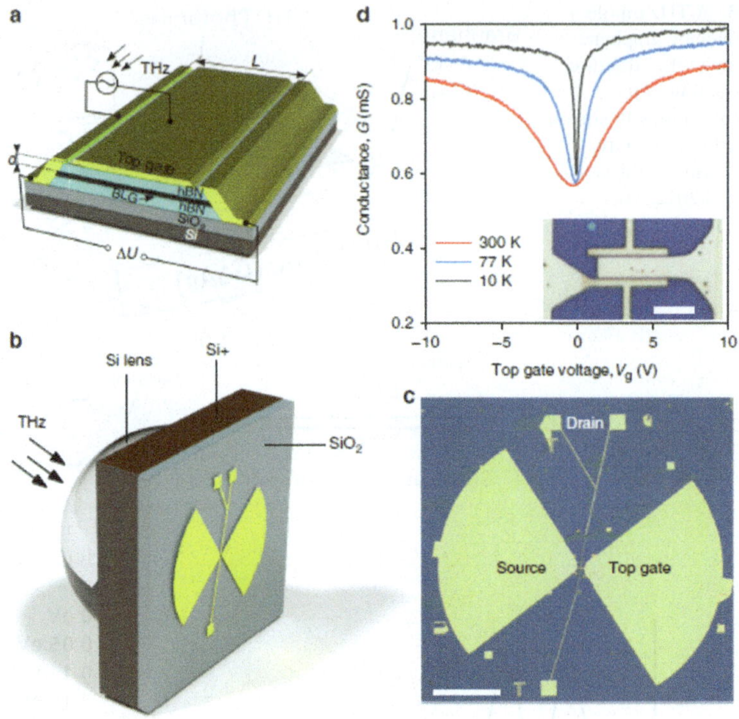

Fig. 10.17 A THz detector based on a bi-layer graphene field-effect transistor: (**a**) and (**b**) views of the schematics; (**c**) picture of the photodetector (scale bar is 0.2 mm); (**d**) conductance versus the gate-source voltage. (Reprinted from [40], 2018 Creative Common License)

solution has been proven to enhance the life of graphene plasmons. The incident radiation modulates the gate-channel voltage, giving raise to the plasmon waves. To increase efficiency, the geometry is designed so that the FET channel behaves as a high-quality plasmonic Fabry-Perot cavities, that strongly enhance the field strength. Finally, the proposed design also provides a rectification of the signal into a DC one.

Unlike all the other design analyzed in this Section, the latter one has been fully realized and characterized. First of all, experimental evidence of the excitation of plasmon resonances have been demonstrated. As for the detector performance analysis, Fig. 10.18 shows the measured values of its responsivity (Volt/Watt) at a resonance frequency of 0.13 THz, as a function of the gate voltage, for three different temperature values. As expected, the most promising values of the responsivity are found for cryogenic temperature values.

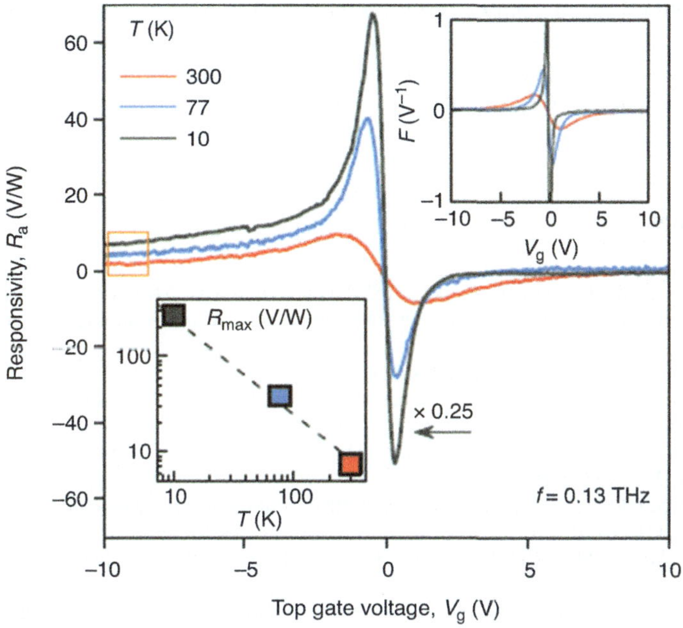

Fig. 10.18 Measured values of the responsitivity of the detector in Fig. 10.17 as a function of the gate voltage, at a resonance frequency of 0.13 THz, for three different values of temperature. Upper inset: FET factor. Lower inset maximum value of responsivity versus temperature. (Reprinted from [40], 2018 Creative Common License)

10.5 Conclusions

We have reviewed the state of the art of graphene resonant antennas, discussing the features, advantages and drawbacks of the most advanced designs and first real-world realizations. All the analyzed antennas exploit the resonances of the surface plasmon polaritons, that are slow surface waves propagating along carbon nanostructures.

The above mechanism has been theoretically demonstrated and experimentally observed and it is now regarded as one of the most promising properties of nanocarbon materials for opening the era of THz nano-antennas with technologically affordable lengths. The other outstanding characteristic of such material is its excellent tunability, by means of several "tuning" parameters, such as electrical and magnetic field and/or doping. These two features allow targeting the two main goals in view of the practical realization of THz antennas: miniaturization and reconfigurability.

Although advanced designs and promising prototypes are available, many challenges have still to be faced to move from design to realization, including theoretical limits (such as the trade-off between radiation efficiency and miniaturization), and fabrication issues (such as stability and reliability of graphene production).

References

1. Siegel PH (2002) Terahertz technology. IEEE Trans Microwave Theory Tech 50:910–928
2. Ferguson B, Zhang XC (2002) Materials for terahertz science and technology. Nature Materials 1:26–33
3. Kiyomi S (2005) Terahertz optoelectronics. Springer, Berlin
4. Dhillon SS et al (2017) The 2017 terahertz science and technology roadmap. J Phys D: Appl Phys 50:043001
5. Neil GR (2014) Accelerator sources for THz science: A Review. J Infrared Millimeter and Terahertz Waves 35:5–16
6. Ganichev SD, Prettl W (2006) Intense terahertz excitation of semiconductors. Oxford University Press, Oxford
7. Lee GM, Wanke MC (2007) Searching for a solid-state terahertz technology. Science 316: 64–65
8. Nagatsuma T, Ducournau G, Renaud CC (2016) Advances in terahertz communications accelerated by photonics. Nature Photonics 10:371–379
9. Vitello MS, Scalari G, Williams B, De Natale P (2015) Quantum cascade lasers: 20 years of challenges. Optics Express 23:5167–5182
10. Rotter S, Gigan S (2017) Light fields in complex media: Mesoscopic scattering meets wave control. Rev Mod Phys 89:015005
11. Xiang ZL, Ashhab S, You JQ, Nori F (2013) Hybrid quantum circuits: Superconducting circuits interacting with other quantum systems. Rev Mod Phys 85:623–653
12. Ritsch H, Domokos P, Brennecke F, Esslinger T (2013) Cold atoms in cavity-generated dynamical optical potentials. Rev Mod Phys 85:553–601
13. Slepyan GY, Boag A, Mordachev V, Sinkevich E, Maksimenko S, Kuzhir P, Miano G, Portnoi ME, Maffucci A (2017) Anomalous electromagnetic coupling via entanglement at the nanoscale. New J Phys 19:023014
14. Gabelli J, Fèeve G, Berroir J-M, Placais B (2012) A coherent RC circuit. Rep Prog Phys 75:126504
15. Kavokin AV, Shelykh IA, Taylor T, Glazov MM (2012) Vertical cavity surface emitting terahertz laser. Phys Rev B 108:197401
16. Hartmann RR, Kono J, Portnoi ME (2014) Terahertz science and technology of carbon nanomaterials. Nanotechnology 25:322001
17. Avouris P, Chen Z, Perebeinos V (2007) Carbon-based electronics. Nature Nanotechn 2: 605–615
18. Neto AHC, Guinea F, Peres NMR, Novoselov KS, Geim AK (2009) The electronic properties of graphene. Rev Mod Phys 81:109–162
19. Hanson GW (2005) Fundamental transmitting properties of carbon nanotube antennas. IEEE Trans. Antennas Prop. 53:3426–3435
20. Li H, Xu C, Srivastava N, Banerjee K (2009) Carbon nanomaterials for next-generation interconnects and passives: Physics, status, and prospects. IEEE Trans. Electron Dev. 56:1799–1821
21. Chen X, Akinwande D, Lee K-J, Close GF, Yasuda S, Paul BC, Fujita S, Kong J, Wong HSP (2010) Fully integrated graphene and carbon nanotube interconnects for gigahertz high-speed CMOS electronics. IEEE Trans. Electr. Devices 57:3137–3143

22. Valitova I, Amato M, Mahvash F, Cantele G, Maffucci A, Santato C, Martel R, Cicoira F (2013) Carbon nanotube electrodes in organic transistors. Nanoscale 5:4638–4646
23. Todri-Sanial A, Dijon J, Maffucci A (2016) Carbon nanotubes for interconnects: Process, design and applications. Springer, Cham, The Netherlands
24. Li B et al (2010) All-carbon electronic devices fabricated by directly grown single-walled carbon nanotubes on reduced graphene oxide electrodes. Adv Materials 22:3058–3061
25. Srivastava A, Marulanda JM, Xu Y, Sharma A (2015) Carbon-based electronics: transistors and interconnects at the nanoscale. Pan Stanford Publishing, Singapore
26. Morris J (2018) Nanopackaging, Nanotechnologies and Electronics Packaging. Springer International Publishing, Cham
27. Bonaccorso F, Sun Z, Hasan T, Ferrari AC (2010) Graphene photonics and optoelectronics. Nature Photonics 4:611–622
28. Zalevsky Z, Abdulhalim I (2014) Integrated nanophotonic devices. Elsevier B.V, Oxford
29. García de Abajo FJ (2013) Applied physics. Graphene nanophotonics. Science 339:917–918
30. Kuzhir P, Paddubskaya A, Valynets NI, Batrakov KG (2017) Main principles of passive devices based on graphene and carbon films in microwave—THz frequency range. J Nanophotonics 11:032504
31. Batrakov K, Kibis OV, Kuzhir P, Rosenau da Costa M, Portnoi ME (2010) Terahertz processes in carbon nanotubes. J Nanophotonics 4:041665
32. Correas-Serrano D, Gomez-Diaz JS, Perruisseau-Carrier J, Alvarez-Melcon A (2014) Graphene-based plasmonic tunable low-pass filters in the terahertz band. IEEE Trans Nanotechn 13:1145–1153
33. Forestiere C, Maffucci A, Miano G (2010) Hydrodynamic model for the signal propagation along carbon nanotubes. Journal of Nanophotonics 4:041695. /1-20
34. Brun C, Wei TC, Franck P, Chong YC, Congxiang L, Leong CW, Tan D, Kang TB, Coquet P, Baillargeat D (2015) Carbon nanostructures dedicated to millimeter-wave to THz interconnects. IEEE Trans Terahertz Sci Techn 5:383–390
35. Burke PJ (2004) AC performance of nanoelectronics: towards a ballistic THz nanotube transistor. Solid-State Electronics 48:1981–1986
36. Shuba MV, Slepyan GY, Maksimenko SA, Thomsen C, Lakhtakia A (2009) Theory of multiwall carbon nanotubes as waveguides and antennas in the infrared and the visible regimes. Phys Rev B 79:155403
37. Carrasco E, Perruisseau-Carrier J (2013) Reflect array antenna at terahertz using graphene. IEEE Antennas Wireless Propag Lett 12:253–256
38. Kibis OV, Rosenau da Costa M, Portnoi ME (2007) Generation of terahertz radiation by hot electrons in carbon nanotubes. Nano Lett 7:3414–3417
39. Vicarelli L et al (2012) Graphene field-effect transistors as room-temperature terahertz detectors. Nature Materials 11:1–7
40. Bandurin DA et al (2018) Resonant terahertz detection using graphene plasmons. Nature Communications 9:5392
41. Novoselov KS, Geim AK, Morozov SV, Jiang D, Zhang Y, Dubonos SV, Grigorieva IV, Firsov AA (2004) Electric field effect in atomically thin carbon films. Science 306:666–669
42. Saito R, Dresselhaus G, Dresselhaus MS (2004) Physical properties of carbon nanotubes. Imperial College Press, Singapore
43. Miano G, Forestiere C, Maffucci A, Maksimenko SA, Slepyan GY (2011) Signal propagation in single wall carbon nanotubes of arbitrary chirality. IEEE Trans Nanotechn 10:135–149
44. Wakabayashi K, Sasaki K, Nakanishi T, Enoki T (2010) Electronic states of graphene nanoribbons and analytical solutions. Sci Technol Adv Mater 1:054504
45. Maffucci A, Miano G (2013) Number of conducting channels for armchair and zig-zag graphene nanoribbon interconnects. IEEE Trans Nanotechn 12:817–823
46. Slepyan GY, Shuba MV, Maksimenko SA, Lakhtakia A (2006) Theory of optical scattering by achiral carbon nanotubes and their potential as optical nanoantennas. Phys Rev B 73:195416

47. Slepyan GY, Maksimenko SA, Lakhtakia A, Yevtushenko O, Gusakov AV (1999) Electrodynamics of carbon nanotubes: Dynamics conductivity, impedance boundary conditions, and surface wave propagation. Phys Rev B 60:17136
48. Hanson GW (2008) Dyadic Green's functions and guided surface waves for a surface conductivity model of graphene. J Appl Phys 103:64302
49. Burke PJ, Li S, Yu Z (2006) Quantitative theory of nanowire and nanotube antenna performance. IEEE Trans Nanotechn 5:314–334
50. Llatser I, Kremers C, Cabellos-Aparicio A, Jornet JM, Alarco E, Chigrin DN (2012) Graphene-based nano-patch antenna for terahertz radiation. Photonics and Nanostructures - Fundam. Appl. 10:353–358
51. Slepyan G, Shuba MV, Maksimenko SA, Thomsen C, Lakhtakia A (2010) Terahertz conductivity peak in composite materials containing carbon nanotubes: Theory and interpretation of experiment. Phys Rev B 81:205423
52. Bommeli F, Degiorgi L, Wachter P, Bacsa WS, de Heer WA, Forro L (1996) Evidence of anisotropic metallic behaviour in the optical properties of carbon nanotubes. Solid State Comm 9:513–517
53. Ugawa A, Rinzler AG, Tanner DB (1999) Far-infrared gaps in single-wall carbon nanotubes. Phys Rev B 60:R11305–R11308
54. Borondics F, Kamarás K, Nikolou M, Tanner DB, Chen ZH, Rinzler AG (2006) Charge dynamics in transparent single-walled carbon nanotube films, from optical transmission measurements. Phys Rev B 74:045431
55. Kampfrath T, von Volkmann K, Aguirre CM, Desjardins P, Martel R, Krenz M, Frischkorn C, Wolf M, Perfetti L (2008) Mechanism of the far-infrared absorption of carbon-nanotube films. Phys Rev Lett 101:267403
56. Akima N et al (2006) Strong anisotropy in the far-infrared absorption spectra of stretch-aligned single-walled carbon nanotubes. Adv Mater 18:1166–1169
57. Waterman PC, Truell R (1961) Multiple scattering of waves. J Math Phys 2:512–537
58. Lakhtakia A (1993) Application of the Waterman-Truell approach for chiral composites. Int J Electron 75:1243–1249
59. Shuba MV, Paddubskaya AG, Plyushch AO, Kuzhir PP, Slepyan GY, Maksimenko SA, Ksenevich VK, Buka P, Seliuta D, Kasalynas I, Macutkevic J, Valusis G, Thomsen C, Lakhtakia A (2012) Experimental evidence of localized plasmon resonance in composite materials containing single-wall carbon nanotubes. Phys Rev B 85:165435
60. Zhang Q, Hároz EH, Jin Z, Ren L, Wang X, Arvidson RS, Lüttge A, Kono J (2013) Plasmonic nature of the terahertz conductivity peak in single-wall carbon nanotubes. Nano Lett 13:5991–5996
61. Nemilentsau AM, Shuba MV, Slepyan GY, Kuzhir PP, Maksimenko SA, D'yachkov PN, Lakhtakia A (2010) Substitutional doping of carbon nanotubes to control their electromagnetic characteristics, Phys Rev B 82: 235424.
62. Shuba MV, Paddubskaya AG, Kuzhir PP, Slepyan GY, Seliuta D, Kasalynas I, Valusis G, Lakhtakia A (2012) Effects of inclusion dimensions and p-type doping in the terahertz spectra of composite materials containing bundles of single-wall carbon nanotubes. J Nanophotonics 6:061707
63. Chiariello AG, Maffucci A, Miano GA (2013) Circuit models of carbon-based interconnects for nanopackaging. IEEE Trans Components, Packaging and Manufacturing 3:1926–1937
64. Slepyan GY, Boag A, Mordachev V, Sinkevich E, Maksimenko S, Kuzhir P, Miano G, Portnoi ME, Maffucci A (2015) Nanoscale electromagnetic compatibility: quantum coupling and matching in nanocircuits. IEEE Trans. on Electromagn Compatibility 57:1645–1654
65. Fichtner N, Zhou X, Russer P (2008) Investigation of carbon nanotube antennas using thin wire integral equations. Adv. Radio Sci. 6:209–211
66. Maffucci A (2018) Carbon interconnects. In: Morris J (ed) Nanopackaging, nanotechnologies and electronics packaging. Springer International Publishing, Cham, pp 725–774
67. Woessner A et al (2014) Highly confined low-loss plasmons in graphene–boron nitride heterostructures. Nature Mater. 14:421–425

68. Yasir M, Savi P, Bistarelli S, Cataldo A, Bozzi M, Perregrini L, Bellucci S (2017) A planar antenna with voltage-controlled frequency tuning based on few-layer graphene. IEEE Antenn. Wireless Propagat Lett 16:2380–2383
69. Llatser I, Kremers C, Chigrin DN, Jornet JM, Lemme MC, Cabellos-Aparicio A, Alarcon E (2012) Radiation characteristics of tunable graphennas in the terahertz band. Radioengineering 21:946–953
70. Li W et al (2014) Ultrafast all-optical graphene modulator. Nano Lett 14:955–959
71. Batrakov KG, Saroka VA, Maksimenko SA, Thomsen C (2012) Plasmon polariton slowing down in graphene structures. J Nanophotonics 6:061719
72. Ryzhii V, Dubinov AA, Aleshkin VY, Ryzhii M, Otsuji T (2013) Injection terahertz laser using the resonant inter-layer radiative transitions in double-graphene-layer structure. Appl Phys Lett 103:10–14
73. Correas-Serrano D, Gomez-Diaz JS, Alu A, Alvarez-Melcon A (2015) Electrically and magnetically biased graphene-based cylindrical waveguides: Analysis and applications as reconfigurable antennas. IEEE Trans. Terahertz Sci. Technol. 5:951–960
74. Zhu B, Ren G, Gao Y, Yang Y, Lian Y, Jian S (2014) Graphene-coated tapered nanowire infrared probe: a comparison with metal-coated probes. Optics Express 22:24096–24103
75. Tamagnone M, Gómez-Diaz JS, Mosig JR, Perruisseau-Carrier J (2012) Reconfigurable terahertz plasmonic antenna concept using a graphene stack. Appl Phys Lett 101:214102

Chapter 11
Terahertz Applications of Non-Simply-Connected and Helical Nanostructures

Thomas P. Collier, Vasil A. Saroka, Charles A. Downing, Arseny M. Alexeev, Richard R. Hartmann, and Mikhail E. Portnoi

Abstract We outline a range of proposals on using non-simply-connected and helical nanostructures for terahertz device implementations. We show that an Aharonov-Bohm quantum ring system and a double-gated quantum ring system both permit control over the polarization properties of the associated terahertz radiation. In addition, we review the superlattice properties of a nanohelix in external electric fields, which reveals negative differential conductance and photogalvanic effects. We present several schemes utilizing carbon nanotubes, including population inversion proposals for both quasi-metallic nanotubes (via generation of optically active hot electrons by an electric field), and metallic nanotubes (by optical excitation across a magnetic field-induced band gap).

Keywords Quantum rings · Nanohelices · Carbon nanotubes · Graphene · Nano-electronics · Nano-electromagnetics · Terahertz range

11.1 Introduction

The aptly named THz gap is a narrow region of the electromagnetic spectrum for which practical and portable technologies lack the ability to produce or detect coherent radiation [1, 2]. Bridging this gap constitutes one of the trickiest problems

T. P. Collier · A. M. Alexeev · M. E. Portnoi (✉)
School of Physics, University of Exeter, Exeter, UK
e-mail: M.E.Portnoi@exeter.ac.uk

V. A. Saroka
Institute for Nuclear Problems, Belarusian State University, Minsk, Belarus

C. A. Downing
Department of Theoretical Physics, Universidad Autonoma de Madrid, Madrid, Spain

R. R. Hartmann
Physics Department, De La Salle University, Manila, Philippines

A. Maffucci, S. A. Maksimenko (eds.), *Fundamental and Applied Nano-Electromagnetics II*, NATO Science for Peace and Security Series B: Physics and Biophysics, https://doi.org/10.1007/978-94-024-1687-9_11

of modern applied physics, and offers potentially diverse device applications spanning from noninvasive biomedical imaging to stand-off detection of plastic explosives [3]. In this chapter, we present several original proposals of non-simply connected and chiral nanostructures as active elements in optoelectronic devices operating in the sought-after THz frequency range [4–13]. These ideas are based on the highly tunable electronic properties which stems from their unusual geometries.

11.2 Quantum Rings

The relatively recent progress in epitaxial growth techniques has led to the burgeoning realm of quantum ring (QR) physics [14]. The fascination with these novel nanostructures originates from the range of exotic quantum phenomena predominantly resulting from the Aharonov-Bohm effect [15, 16]. Piercing a quantum ring by a magnetic flux ϕ yields the single-electron energy spectrum

$$\varepsilon_m(\phi) = (m + \phi/\phi_0)^2 \varepsilon_1(0), \tag{11.1}$$

where $\phi_0 = h/e$ is the flux quanta, and $m \in \mathbb{Z}$ denotes the angular momentum quantum number. For a typical semiconductor ring of radius R and electron effective mass M_e, the energy scale $\varepsilon_1(0) = \hbar^2/2M_eR^2$ lies in the THz regime [17, 18]. The energy spectrum exhibits Aharonov-Bohm oscillations in magnetic flux with a period of ϕ_0, and a half-integer number of flux quanta leads to a degeneracy of the electron energy levels corresponding to m differing by one. This degeneracy can be easily lifted by an electric field E applied in the plane of the ring, which breaks the axial symmetry of the system and mixes states with different angular momenta. As such, it can be shown [4, 5] that the lowest two energy levels become separated by a gap proportional to the electric field strength $\Delta\varepsilon^{(\phi=\phi_0/2)} = eER$. Incidentally, at zero flux the energy separation has a quadratic dependence $\Delta\varepsilon^{(\phi=0)} = \varepsilon_1(0) - (eER)^2/2\varepsilon_1(0)$. A consequence of even relatively mild electric fields $eER{\sim}0.1\varepsilon_1(0)$ is the suppression of Aharonov-Bohm ground state energy oscillations, which is a cause of frustration for device proposals. Other quantities like the ring's dipole moment and polarization-dependent selection rules may prove to be practical in lieu of this.

Consider an electron occupying the nth state Ψ_n of a neutral ring in the presence of a uniform positive background (modelled by $+e$ at the center of the ring). The projection of the dipole moment direction on the lateral E-field is,

$$P_n = eR \int |\Psi_n|^2 \cos\varphi d\varphi, \tag{11.2}$$

where the angular coordinate φ is measured from the horizontal axis defined by the E-field direction. In the absence of such an E-field, each state is characterized by a value of m ($\Psi_m = e^{im\varphi}/\sqrt{2\pi}$) and the corresponding charge density is uniformly spread over the ring. In the presence of a weak field $eER \ll \varepsilon_1(0)$ and zero magnetic field, the $m = 0$ ground state picks up only tiny contributions from wave functions with $m \neq 0$ and the charge distribution remains predominantly uniformly spread. However, at a degeneracy-inducing value of magnetic flux $\phi = \phi_0/2$, the ground state φ-dependence is well described by sin(φ/2), resulting in a shift of the charge density distribution *against* the electric field. Therefore, upon changing flux, the ground state goes from unpolarized to strongly polarized with a dipole moment oscillation period of ϕ_0.

Introducing finite temperature into the model requires considering the non-zero occupations of excited states, as the first excited state picks up a cos(φ/2) dependence at half-flux and compensates for the dipole moment of the ground state. Calculations for the dipole moment, after thermal averaging over all states $<P>$, suggests that while magneto-oscillations of the dipole moment for nano-scale rings are observable in principle, the requirement of temperatures below 2 K prevents portable device applications. The system's inter-level optical transitions, however, are less sensitive to the partial occupation of excited states at finite temperature. The transition dipole matrix element $P_{if}(\theta)$ for linearly polarized radiation incident onto the QR dictates the transition rate between initial (*i*) and final (*f*) states $T_{if}(\theta) \propto |P_{if}(\theta)|^2$, and for our ring takes the form,

$$P_{if}(\theta) = eR \int \Psi_f{}^* \Psi_i \cos(\theta - \varphi)\, d\varphi, \tag{11.3}$$

where θ is the angle between the projection of linearly polarized radiation onto the plane of the QR and the direction of the lateral *E*-field. Away from the degeneracies, transitions between ground and first excited state have no linear polarization. At $\phi = \phi_0/2$, the transition rate with $\theta = \pi/2$ polarization ($T_{01}(\pi/2) \equiv T_\perp$) reaches its maximum possible value, whereas transitions with $\theta = 0$ are forbidden ($T_{01}(0) \equiv T_\parallel = 0$). Such oscillations of the degree of polarization and strong optical anisotropy do not depend on temperature. Magneto-oscillations of both the dipole moment of the ring and the degree of polarization of inter-level transitions are plotted in Fig. 11.1.

An additional degree of control can be achieved by placing a ring in a THz cavity [6]. If the flux piercing the ring is $\phi = \phi_0/2$, we require only a small change in *E*-field to tune the energy levels of the ring into resonance with a single-mode microcavity. Additionally, the QR-microcavity coupling constant (for a linearly polarized cavity mode) is given by

$$\mathcal{G} = -P_{if}(\theta)\sqrt{\frac{\hbar\omega_{MC}{}^4}{2\pi^3 c^3}}, \tag{11.4}$$

where ω_{MC} is the microcavity mode frequency. Hence, contrary to quantum dots, for such a QR system both $\Delta\varepsilon$ and $\mathcal{G}$ can be controlled with external electric and magnetic fields. In Fig. 11.2 we plot the QR ring and microcavity emission spectra,

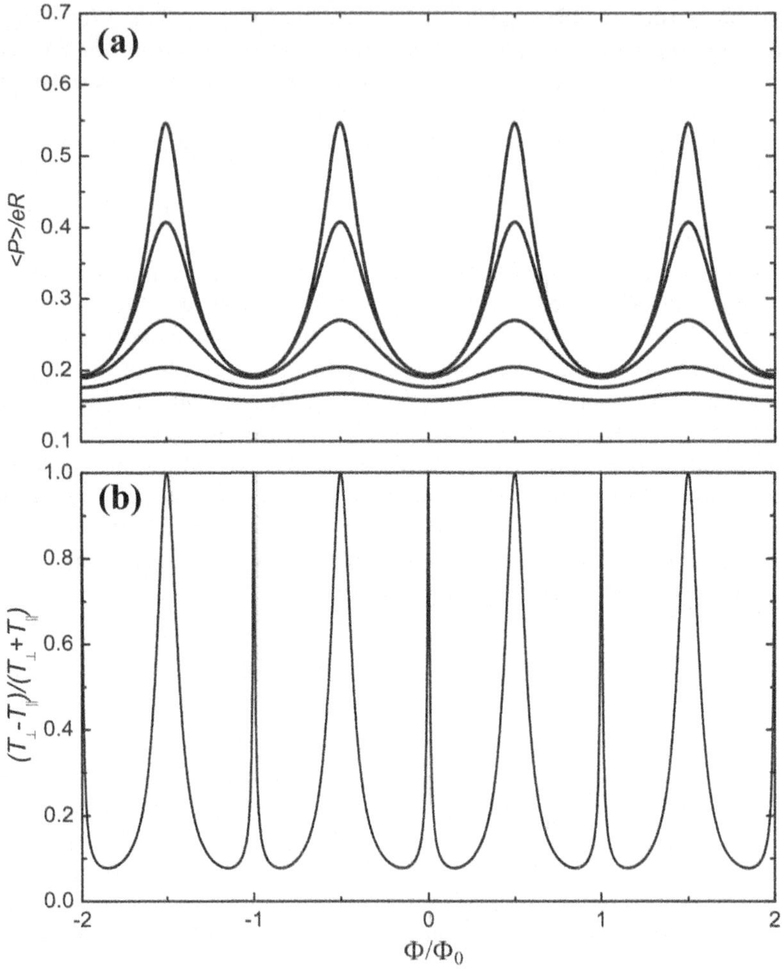

Fig. 11.1 (**a**) Plots magneto-oscillations of the dipole moment of a ring for temperatures ranging from $0.01\varepsilon_1(0)/k_B$ (upper curve) to $0.41\varepsilon_1(0)/k_B$ (lower curve) with increments of $0.1\varepsilon_1(0)/k_B$. (**b**) Shows magneto-oscillations of the degree of polarization for transitions between the two lowest energy levels. The subscripts $\perp$ and $\|$ denote transitions polarized perpendicular and parallel to the electric field respectively. (Adapted from Ref. [4])

$S_{QR}(\omega)$ and $S_{MC}(\omega)$ respectively, for the off-resonance system ($\Delta \neq \hbar\omega_{MC}$) at zero flux and at half-integer flux quanta for different values of E. In conclusion, we have shown that Aharonov-Bohm rings can act as room-temperature polarization-sensitive THz detectors and optical magnetometers, or as a system for population inversion via optical excitation of an electron into the first excited state across the semiconductor band gap with a tunable radiative transition to the ground state. If embedded in a microcavity the system has a highly controllable emission spectrum and can be used as a tunable optical modulator or as a tool for magneto-spectroscopy.

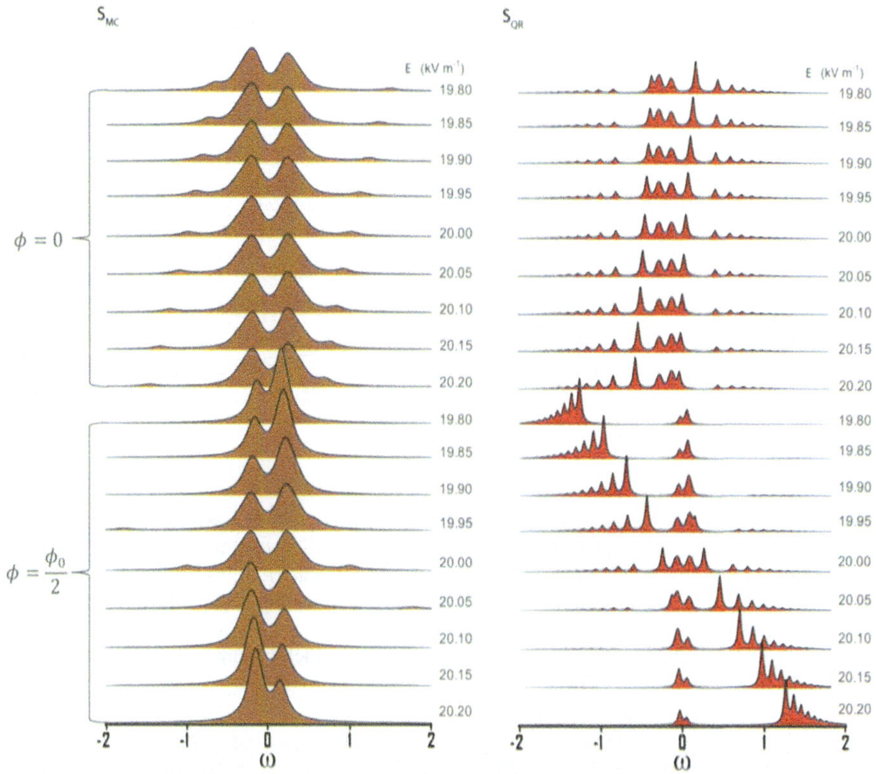

Fig. 11.2 Anticrossing in the emission spectra of the QR-microcavity system at different magnitudes of the lateral electric field, with magnetic flux piercing the ring at $\phi = 0$ (upper plots) or $\phi = \phi_0/2$ (lower plots). The gold left-hand plots show the microcavity emission spectrum whereas red right-hand plots show the direct QR emission spectrum. The resonance case $\Delta = \hbar\omega_{MC}$ corresponds to $E = 20.00\,\mathrm{kVm^{-1}}$ and the emission frequencies normalized by the QR-microcavity coupling constant $\mathcal{G}/\hbar$ are centered around ω_{MC}. (Adapted from Ref. [6])

Exploiting Aharonov-Bohm related properties requires trapping up to one-half of the magnetic flux quantum in the ring annulus, which requires fields of several Tesla for a ~10 nm radius semiconductor ring. We show that the use of prohibitively strong magnetic fields can be avoided if a ring is placed between two lateral gates [7], which induces a double quantum well potential along the ring

$$V(\varphi) = 2\beta \frac{d}{R}(1-\gamma)\cos(\varphi) + \beta\gamma\cos(2\varphi), \tag{11.5}$$

where β is the potential strength from one gate at a distance d from the ring and γ characterizes the relative potential strength of the other gate ($\gamma = 1$ describes equivalent gates). The well parameters and corresponding inter-level separations are highly sensitive to the gate voltages. Our analysis for double-gated rings shows

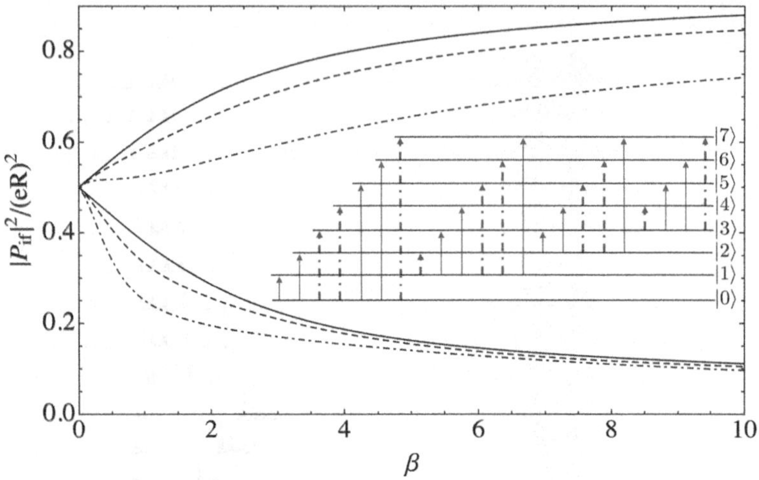

Fig. 11.3 Square of the dimensionless transition dipole matrix element between initial (i) and final (f) single-electron states plotted as a function of β and with $\gamma = 1$ (full line), $\gamma = 1.003$ (dashed line), and $\gamma = 1.006$ (dot-dashed line). The upper branches denote $|P_{01}|^2/(eR)^2$ at $\theta = \pi/2$ and the lower branches denote $|P_{02}|^2/(eR)^2$ at $\theta = 0$. The inset schematically depicts the optical selection rules between the lowest eight energy eigenstates. Full blue arrows are allowed transitions when $\gamma = 1$, the red dot-dashed arrows are forbidden transitions which become allowed when $\gamma \neq 1$. (Adapted from Ref. [7])

that selection rules, caused by linearly polarized excitations for inter-level dipole transitions Eq. (11.3), depend on the polarization vector angle θ with respect to the gates. In Fig. 11.3 we plot the transitions between the ground and first (second) excited state for polarization angles $\theta = \pi/2$ ($\theta = 0$), the inset schematically represents the selection rules for both identical and asymmetrical gate voltages. In striking difference from the conventional symmetric double well potential, the ring geometry permits polarization-dependent transitions between the ground and second excited states, allowing the use of this structure in a three-level lasing scheme.

11.3 Nanohelices

The helical form is prevalent throughout nature, not least of all contributing to the structure of DNA. In condensed matter physics, nanohelices have been successfully realised in semiconductor systems [19], and vast improvements in uniformity and quality of such structures can be seen from the exotic phenomena reported in this geometry, such as the quantum hall effect [20]. In what follows we elucidate how a nanohelix in the presence of an electric field normal to its axis behaves like a superlattice with tunable parameters.

The problem of a QR pierced by a flux and subjected to an in-plane electric field is mathematically equivalent to that of an ideal helix subjected to an electric field normal to its axis [8–10]. The role of the flux is played by the electron momentum along the helix. Compared to Aharonov-Bohm rings, (as with the double-gated QR system) the helix geometry has the advantage of not needing high magnetic fields to yield desirable physics. Consider a nanohelix in an external electric field normal to the helix axis $E_\perp$ as a one-dimensional conductor with total helical length L, radius R, and pitch d, as shown in the inset of Fig. 11.5. The potential energy of an electron along the helical line coordinate s is then

$$V(s) = eE_\perp R \cos(2\pi s/l_0)\,, \tag{11.6}$$

where $l_0 = \sqrt{4\pi^2 R^2 + d^2}$ is the length of a single coil. Clearly, $V(s)$ is periodic with period l_0 significantly larger than the interatomic distance, resulting in typical superlattice behaviour with tuneable electronic properties. The energy of an electron in a field-free helix is $\varepsilon_0(k) = \hbar^2 k^2/2M_e$. For weak electric fields $eE_\perp R \ll \varepsilon_0(2\pi/l_0)$, we can assume $V(s)$ only mixes adjacent states, which gives the spectrum

$$\varepsilon(k) = \frac{1}{2}\left[\varepsilon_0(k) + \varepsilon_0\left(|k| + \frac{2\pi}{l_0}\right)\right] \pm \frac{1}{2}\sqrt{\left[\varepsilon_0(k) + \varepsilon_0\left(|k| + \frac{2\pi}{l_0}\right)\right]^2 + eE_\perp R}. \tag{11.7}$$

Thus, a linearly dependent band gap $\Delta\varepsilon^{(k=\pm\pi/l_0)} = eE_\perp R$ is opened by $E_\perp$ at the first Brillouin zone edge due to the Bragg scattering of electrons from the long-range periodic potential, and can be tuned to the eulogized THz range by experimentally attainable external fields. Neglected in this approximation are the band gaps at other Brillouin zone boundaries (of higher order in $E_\perp$). In the limit of strong field $eE_\perp R \gg \varepsilon_0(2\pi/l_0)$ the electronic spectrum tends towards dispersionless flat bands as the potential approaches the limit of an impenetrable periodic harmonic potential [10]. In Fig. 11.4a we plot, as a function of wave vector along the helix axis (z-axis) k_z, results for the transition rate T_{01} across this energy gap for both linearly polarized light normal to the helix axis and for right-hand circularly polarized light going along it. The result for left-handed polarization is a mirror image of the on-off switching behavior across the center of the Brillouin zone exhibited by the right-handed polarization, which suggests a photogalvanic effect (whereby one may choose the net direction of electrons by shining the appropriate circularly polarized light).

When a longitudinal electric field $E_\parallel$ is also applied to this system [10], such that the semi-classical motion along the z-axis is $k(t) = eE_\parallel t/\hbar$, we can see from the drift velocity v_d plotted as a function of applied field in Fig. 11.4b that beyond a critical field $E_\parallel > E_\tau$ (where $E_\tau = \hbar/ed\tau$ and τ is a phenomenological scattering time) the decreasing drift velocity implies a negative differential conductance (since the current $I \propto v_d$). Upon considering tunneling between the first and ground band, we find that above a certain field $E_\parallel \gtrsim 4E_\tau$ the drift velocity increases once more (corresponding to N-type current-voltage characteristics).

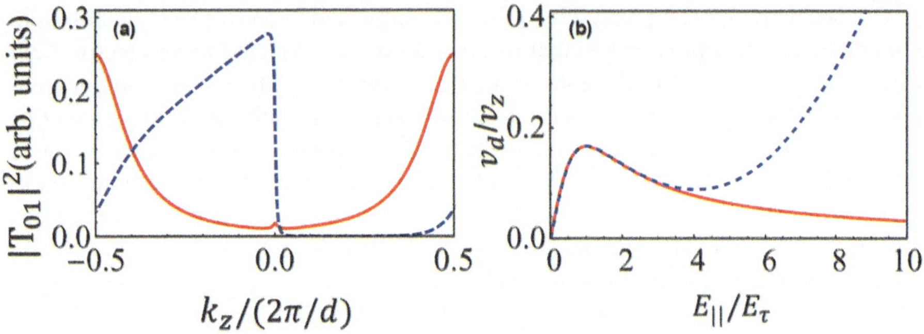

Fig. 11.4 (**a**) Square of the dimensionless transition dipole matrix element between the ground and first excited state in the first Brillouin zone as a function of the dimensionless wave vector k_z. The dashed blue line is for transitions induced by right-handed circularly polarized light propagating along the helix axis whereas the full red line is for linearly polarized light polarized parallel to the helix axis. (**b**) Drift velocity as a function of external longitudinal electric field, without (full red line) and with (dashed blue line) the effect of tunneling from the ground band considered. We plot here for parameters $eE_\perp R = 0.4\varepsilon_0(d/2\pi)$ and $\varepsilon_0(d/2\pi)\tau = 10\hbar$, while v_z is the magnitude of the velocity maxima at the Brillouin zone edge in zero external fields. (Adapted from Ref. [10])

11.4 Carbon Nanotubes

The aforementioned nanohelix-based phenomena will also manifest in specific types of chiral carbon nanotubes (CNT) [8, 9]. As depicted in Fig. 11.5, electron motion in a $(n, 1)$ carbon nanotube corresponds to a de Broglie wave along a helical trajectory. In particular, the inherent regularity of CNT-based helical supperlattices offers an advantage over conventional imperfect heterostures and may contribute to nanoelectronic devices such as Bloch oscillators and quantum cascade lasers. We note that the discussed superlattice behaviour is unique to this chiral CNT, as the energy spectrum of (n, m) nanotubes depends also on the electron angular momentum projected onto the nanotube axis. The transverse electric field then mixes electron states from different subbands with typically different energies, and inevitably this does not noticeably change the electronic energy dispersion for weak electric fields. However, as we will elude to here, other CNT types in external fields are promising structures for THz applications and exhibit a plethora of diverse effects [8, 9, 11–13]. Here, and in what follows, we refer only to the properties of single-wall CNTs.

The heating of the electron gas in a CNT, via the application of an electric field along the nanotube axis, results in a population inversion of optically active states with an energy difference in the THz regime. However, due to the suppression of elastic backscattering mechanisms characteristic of metallic CNTs, in high enough electric fields the charge carriers can be sufficiently accelerated to induce the emission of phonons – which in turn corresponds to a swift current saturation. We show here that a quasi-metallic CNT in the low-bias regime, and with large radius R, can produce spontaneous emission of THz radiation prior to the phonon-emission threshold. The peak-frequency of THz emission is also tuned by the applied voltage.

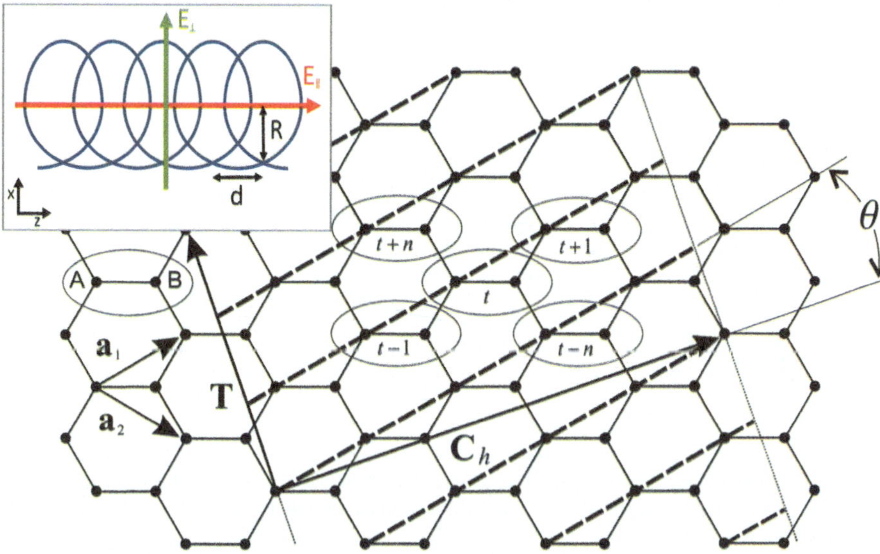

Fig. 11.5 Schematic of a CNT formed from an unrolled graphene. By connecting the head and tail of the Chiral vector C_h we can construct a helical $(n, 1)$ CNT which can be described by transition of an elementary two-atom cell along the dashed helical line; shown here is a (4,1) nanotube. The inset depicts the geometry of the considered helix in external transverse and longitudinal electric fields. (Adapted from Ref. [8, 9], inset adapted from Ref. [10])

The electronic spectrum near the fermi energy $\varepsilon(k) = \pm\hbar v_F|k - k_0|$ is linearly dependent on the wavevector k, where $v_F = 9.8 \times 10^5$m/s is ther Fermi velocity for graphene [21]. Upon applying a voltage V between the ends of the nanotube, the k-state electron distribution is shifted (see Fig. 11.6a) such that optical transitions can occur between the recently occupied states in the conduction bands and emptied states in the valence band. To avoid the detrimental electron-phonon interaction, we require CNTs of length L $\ll l_{ac}$, where $l_{ac} \approx 2\mu$m [22] is the mean-free path for electrons scattering from acoustic phonons. The energy of a ballistic electron picked up along L, $\Delta\varepsilon = eV$, must remain below the high-energy phonon threshold of 0.16 eV [19]. The distribution function for both hot electrons and holes is $f_{e,h}(k) = 1$ for $0 < k - k_0 < \Delta\varepsilon/2\hbar v_F$ and $f_{e,h}(k) = 0$ for $k - k_0 > \Delta\varepsilon/2\hbar v_F$. Optical transitions between the conduction and valence subbands are forbidden for truly metallic armchair (n, n) CNTs. Such transitions are allowed, however, for quasi-metallic $(n - 3p, n)$ CNTs where $p \in \mathbb{Z}^*$ is a non-zero integer. The curvature-induced narrow gap $\varepsilon_g \propto R^{-2}$ can be neglected for large radii CNTs in electric fields sufficiently strong enough to achieve Zener breakdown for electrons tunneling across the gap. The spectral density of spontaneous emission in the dipole approximation is

$$I_\nu = \frac{8\pi e^2\nu}{3c^2}\sum_{i,f} f_e(k_i)\ f_h(k_f)\ |<\Psi_f\ |\ v_z|\Psi_i>|^2\delta\left(\varepsilon_i - \varepsilon_f - \hbar\nu\right), \tag{11.8}$$

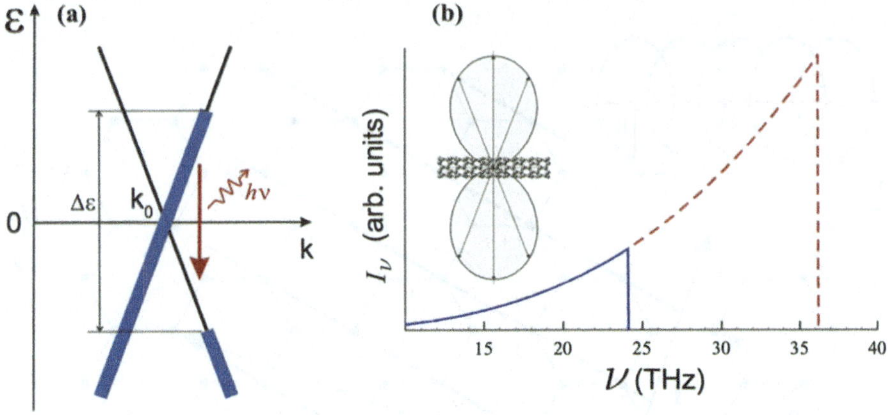

Fig. 11.6 (**a**) Schematic of THz radiation by hot carriers in quasi-metallic CNTs. (**b**) Spectral density of spontaneous emission plotted as a function of photon frequency. The solid line is for $V = 0.1\text{V}$ and the dashed line is for $V = 0.15\text{V}$. The inset depicts the direction of the radiation. (Adapted from Ref. [11])

and represents the rate of change, with respect to photon frequency ν, of the probability of optical transitions per unit time in the interval $(\nu, \nu + d\nu)$. Let us consider a zigzag $(3p, 0)$ CNT. We can express the matrix element of the velocity operator $\boldsymbol{v}_z$ between the lowest conduction and highest valence subbands via

$$< \Psi_f \mid \boldsymbol{v}_z \mid \Psi_i >= \frac{a_{C-C}}{8\hbar} \left(\varepsilon_i - \varepsilon_f\right) \delta_{k_f, k_i}, \qquad (11.9)$$

where $a_{C-C} = 1.42\mathring{A}$ is the nearest neighbour separation between two carbon atoms. Substituting Eq. (11.9) in Eq. (11.8) yields

$$I_\nu = L\text{f}_e \left(\pi \nu / v_f\right) \text{f}_h \left(\pi \nu / v_f\right) \frac{\pi^2 e^2 a_{C-C}{}^2 \nu^3}{6c^3 \hbar v_f}. \qquad (11.10)$$

Figure 11.6b plots I_ν for two different voltages, and shows the strong voltage dependence of the spectral density of THz emission. The inset of Fig. 11.6b depicts the direction of the radiation, and is due to the forbidden optical transitions of light polarized perpendicular to the nanotube axis.

As mentioned, the strictly forbidden optical transitions between the top valence and lowest conduction subbands in a (n, n) armchair CNT prevents its use in proposals based on the inversion of population of optically active states. However, the application of a magnetic field along the CNT axis results in both the opening of a band gap (in the THz spectrum) and altered selection rules, where the previosuly forbidden optical transitions become allowed due to the symmetry-breaking magnetic field. The band gap grows linearly with the field for experimentally reasonable values. For armchair nanotubes the squared velocity operator matrix element between the states at the edge of the magnetically opened gap can be succinctly expressed as

$$\left|< \Psi_n{}^v \mid \boldsymbol{v}_z \mid \Psi_i{}^c >\right|^2 = \frac{4}{3}\left[1 - \frac{1}{4}\cos^2\left(\phi\,\pi/\phi_0 n\right)\right] v_f{}^2 \tag{11.11}$$

where ϕ and ϕ_0 retain their definitions of enclosed flux and flux quanta respectively, and $\phi \le \phi_0/2$ since (analagous to QRs) the order of the subbands is altered for $\phi > \phi_0/2$. In Fig. 11.7 we plot the energy dispersion and matrix element of the dipole optical transitions polarized parallel to the nanotube axis for a (10, 10) CNT with and

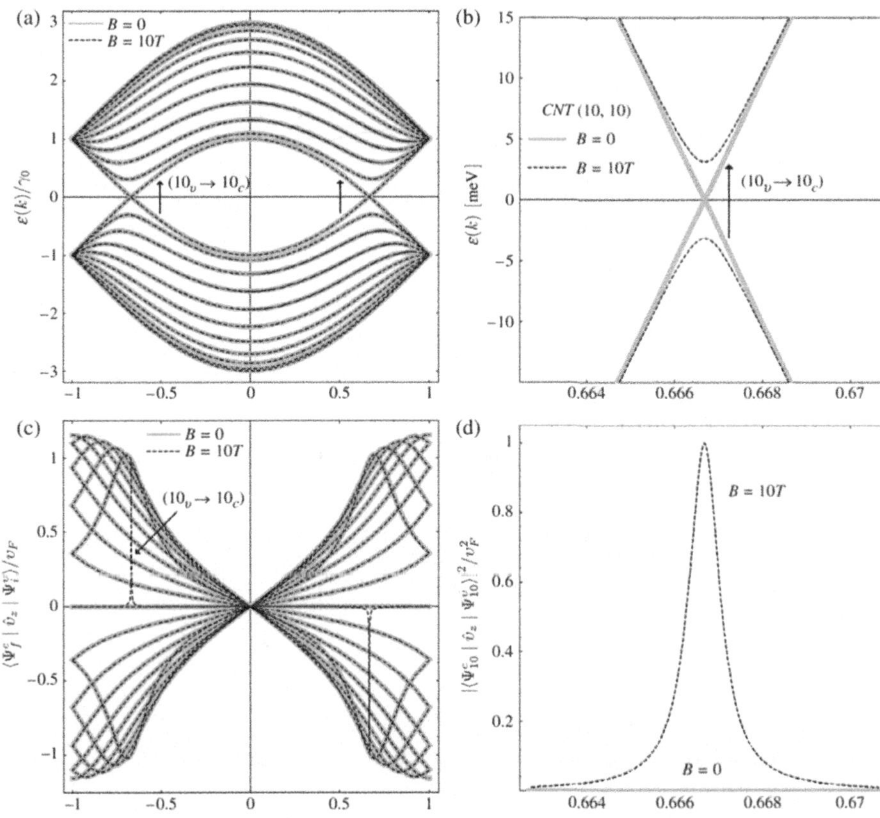

Fig. 11.7 (**a**) Band diagram for a (10, 10) CNT in the first Brillouin zone, with and without magnetic field applied along the CNT axis, with T the crystal period along the nanotube. (**b**) Detailed view of the opened gap between the top valence (10_v) and lowest conduction (10_c) subbands. (**c**) Dipole optical transitions matrix elements, with and without magnetic field, for light polarized along the CNT axis. The only change with magnetic field is now the existence of a narrow peak associated with the previously forbidden ($10_v \to 10_c$) transition. (**d**) Dependence of the squared dipole matrix for this ($10_v \to 10_c$) transition. (Adapted from Ref. [12])

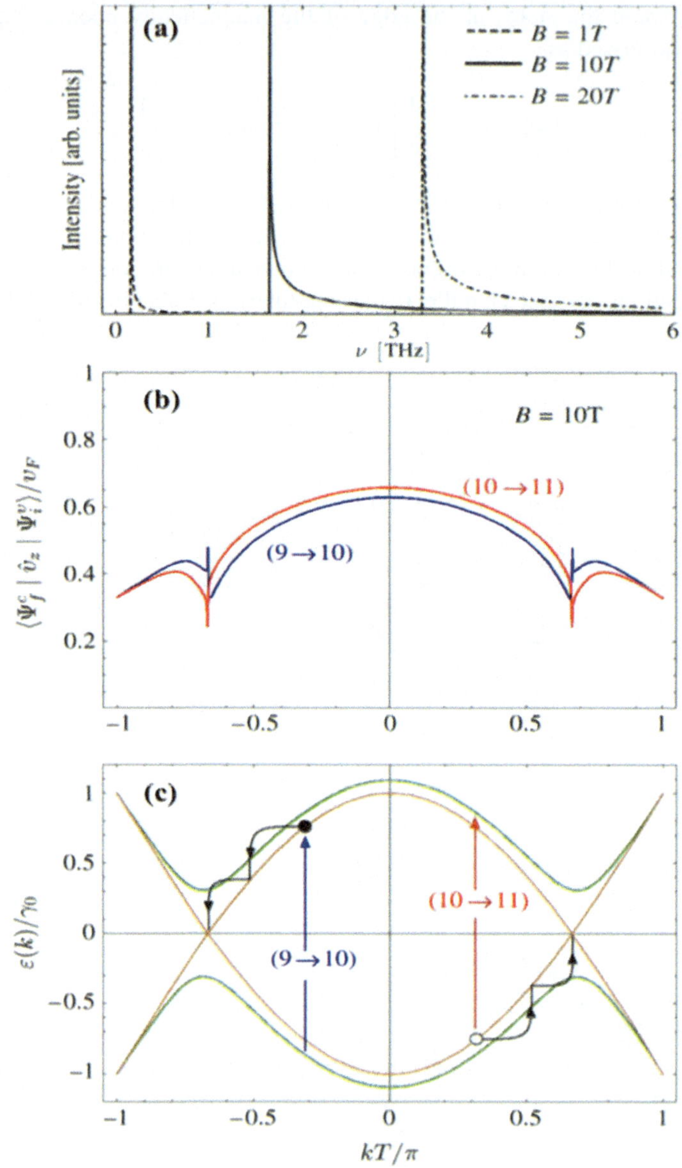

Fig. 11.8 (**a**) Absorption spectra for a (10, 10) CNT for three values of magnetic field, the absorption intensity is proportional to the product of the square of velocity operator matrix element and the joint density of states. (**b–c**) A scheme for achieving population inversion in an armchair CNT in a magnetic field: (**b**) dipole optical matrix elements for transitions polarized normal to the nanotube axis; (**c**) depiction of the creation of photoexcited carriers and the non-radiative thermalization for energy subbands closest to the Fermi energy. (Adapted from Ref. [12])

without an external magnetic directed along the nanotube. The energy spectrum near the bottom or top of the magnetically induced band gap is parabolic with respect to the wavevector k along the nanotube axis, as seen in Fig. 11.7b. Consequently, such a dispersion leads to a Van Hove singularity in the reduced density of states. The appearance of this singularity results in a very sharp absorption maximum near the band edge (Fig. 11.8a), which yields an exceptionally sensitive dependence of the photocurrent on the photon frequency.

The Van Hove singularity can conversely be used to produce a very narrow emission line, with a peak frequency controlled by the magnetic field. Here, a population inversion can now be accomplished in metallic armchair nanotubes by pumping with light polarized perpendicular to the CNT axis, as depicted in Fig. 11.8b and c.

11.5 Conclusions

In this chapter, we have discussed several proposals of using ring- and helix-like nanostructures, as well as carbon nanotubes, to address problems in THz optoelectronics. We have shown that the high tunability of the electronic properties of these structures in external fields stems from their non-simply-connected geometries, and we have demonstrated how corresponding phenomena may be exploited for device applications.

Acknowledgements We would like to acknowledge financial support from the Engineering and Physical Sciences Research Council (EPSRC) of the United Kingdom, via the EPSRC Centre for Doctoral Training in Metamaterials XM2 (Grant No. EP/L015331/1). The work was also financially supported by the EU H2020 RISE project CoExAN (Grant No. H2020-644076) and by the Government of the Russian Federation through the ITMO Fellowship and Professorship Program.

References

1. Siegel PH (2002) IEEE Trans Microwave Theory Technol 50:910
2. Ferguson B, Zhag XC (2002) Nat Mater 1:26
3. Tonouchi M (2007) Nat Photonics 1:97
4. Alexeev AM, Portnoi ME (2012) Phys Rev B 85:245419
5. Alexeev AM, Portnoi ME (2012) Phys Status Solidi C 9:1309
6. Alexeev AM, Shelykh IA, Portnoi ME (2013) Phys Rev B 88:085429
7. Collier TP, Saroka VA, Portnoi ME (2017) Phys Rev B 96:235430
8. Kibis OV, Malevannyy SV, Huggett L, Parfitt DGW, Portnoi ME (2005) Electromagnetics 25:425
9. Kibis OV, Parfitt DGW, Portnoi ME (2005) Phys Rev B 71:035411
10. Downing CA, Robinson MG, Portnoi ME (2016) Phys Rev B 94:155306
11. Portnoi ME, Kibis OV, da Costa MR (2008) Superlattice Microst 43:399
12. Portnoi ME, da Costa MR, Kibis OV, Shelykh IA (2009) Int J Mod Phys B 23:2846

13. Hartmann RR, Kono J, Portnoi ME (2014) Nanotechnology 25:322001
14. (2018) Physics of quantum rings (2nd ed). Nanoscience and technology, by V. Fomin. Springer, Cham
15. Ehrenberg W, Siday RE (1949) Proc Phys Soc London, Sect B 62:8
16. Aharonov Y, Bohm D (1959) Phys Rev 115:485
17. Lorke A, Luyken RJ, Govorov AO, Kotthaus JP, Garcia JM, Petroff PM (2000) Phys Rev Lett 84:2223
18. Ribeiro E, Govorov AO, Carvalho W Jr, Medeiros-Riberio G (2004) Phys Rev Lett 92:126402
19. Ren Z, Gao P–X (2014) Nanoscale 6:9366
20. Vorobyova JS, Vorobes AB, Prinz VY, Toropov AI, Maude DK (2015) Nano Lett 15:1673
21. *Physical properties of carbon nanotubes*, by R. Saito, G. Dresselhaus, M. S. Dresselhaus (Imperial College Press, London, 1998)
22. Park J-Y, Rosenblatt S, Yaish Y, Sazanova V, Üstünel H, Braig S, Arias TA, Brouwer PW, McEuen PL (2004) Nano Lett 4:517

The manufacturer's authorised representative in the EU is Springer Nature Customer Service Centre GmbH, Europaplatz 3, 69115 Heidelberg, Germany. If you have any concerns regarding our products, please contact ProductSafety@springernature.com

Printed and bound by CPI Group (UK) Ltd, Croydon, CR0 4YY
23/07/2026
02175025-0002